Louis Agassiz

Geological Sketches

Louis Agassiz

Geological Sketches

ISBN/EAN: 9783337012977

Printed in Europe, USA, Canada, Australia, Japan

Cover: Foto ©berggeist007 / pixelio.de

More available books at **www.hansebooks.com**

GEOLOGICAL SKETCHES.

BY

L. AGASSIZ.

BOSTON:
JAMES R. OSGOOD AND COMPANY,
LATE TICKNOR & FIELDS, AND FIELDS, OSGOOD, & CO.
1873.

PREFACE.

THE articles collected in this volume, like those formerly published under the title of "Methods of Study," were originally prepared from notes of extemporaneous lectures, and first appeared in the pages of the Atlantic Monthly. They still retain something of the familiarity induced by the personal relation of a lecturer to his audience, so different from the more distant one of the author to his reading public. They must indeed be considered as familiar talks on scientific subjects rather than as scientific papers.

The three concluding chapters on Glaciers are introduced with special reference to their geological significance, and will be followed in a future volume by a number of articles showing the former extension of glaciers on this continent as well as in Europe, and giving at some length the history of their retreat. This will naturally lead to a discussion of all the facts connected with the ice period, the erratic boulders, the drift, the formation of river systems, ancient lake and river terraces,— all the phenomena, in short, resulting from the former presence of such immense masses of ice and their subsequent disappearance. These questions have been chiefly studied on the European continent, where the broken character of the country, intersected in every direction by mountain chains, presents numerous centres of dispersion for glaciers. Owing to the extensive land surfaces on this continent, the same set of facts presents quite a different aspect here and in the Old World; and I hope that the facilities I have had for tracing the glacial phenomena in America may enable me to throw some new light on this subject.

L. AGASSIZ.

CAMBRIDGE, November 29th, 1865.

CONTENTS.

		PAGE
I.	AMERICA THE OLD WORLD	1
II.	THE SILURIAN BEACH	29
III.	THE FERN FORESTS OF THE CARBONIFEROUS PERIOD	64
IV.	MOUNTAINS AND THEIR ORIGIN	94
V.	THE GROWTH OF CONTINENTS	120
VI.	THE GEOLOGICAL MIDDLE AGE	148
VII.	THE TERTIARY AGE, AND ITS CHARACTERISTIC ANIMALS	181
VIII.	THE FORMATION OF GLACIERS	208
IX.	INTERNAL STRUCTURE AND PROGRESSION OF GLACIERS	233
X.	EXTERNAL APPEARANCE OF GLACIERS	283

I.

AMERICA THE OLD WORLD.

FIRST-BORN among the Continents, though so much later in culture and civilization than some of more recent birth, America, so far as her physical history is concerned, has been falsely denominated the *New World.* Hers was the first dry land lifted out of the waters, hers the first shore washed by the ocean that enveloped all the earth beside; and while Europe was represented only by islands rising here and there above the sea, America already stretched an unbroken line of land from Nova Scotia to the Far West.*

In the present state of our knowledge, our conclusions respecting the beginning of the earth's history, the way in which it took form and shape as a distinct, separate planet, must, of course, be very vague and hypothetical. Yet the progress

---

* It would be inexpedient to encumber these pages with references to all the authorities on which such geological results rest. They are drawn from the various State Surveys, including that of the mineral lands of Lake Superior, in which the early rise of the American Continent is for the first time affirmed, and other more general works on American geology.

of science is so rapidly reconstructing the past that we may hope to solve even this problem; and to one who looks upon man's appearance upon the earth as the crowning work in a succession of creative acts, all of which have had relation to his coming in the end, it will not seem strange that he should at last be allowed to understand a history which was but the introduction to his own existence. It is my belief that not only the future, but the past also, is the inheritance of man, and that we shall yet conquer our lost birthright.

Even now our knowledge carries us far enough to warrant the assertion that there was a time when our earth was in a state of igneous fusion, when no ocean bathed it and no atmosphere surrounded it, when no wind blew over it and no rain fell upon it, but an intense heat held all its materials in solution. In those days the rocks which are now the very bones and sinews of our mother Earth — her granites, her porphyries, her basalts, her syenites — were melted into a liquid mass. As I am writing for the unscientific reader, who may not be familiar with the facts through which these inferences have been reached, I will answer here a question which, were we talking together, he might naturally ask in a somewhat skeptical tone. How do you know that this state of things ever existed, and, supposing that the solid materials of which our earth consists were ever in a liquid condition, what right have you to infer that this condition was caused by the action of heat upon them? I answer, Because it is acting upon them still; because the earth we tread is but a thin crust floating on a liquid sea of molten materials; because the agencies that were at work then are at work now, and the present is the logical sequence of the past. From Artesian wells, from mines, from geysers, from hot springs, a mass of facts has been collected, proving incontestably the heated condition of all substances at a certain depth below the earth's surface; and if we need more positive evidence, we have it in the fiery eruptions that even now bear fearful testimony to the molten ocean seething within the globe and forcing its way out from time to time. The modern progress of Geology has led us by successive and perfectly connected steps back to a time when what is now only an occasional and rare phenomenon was the normal condition of our earth; when those internal fires were enclosed in an envelop so thin that it opposed but little resistance to their frequent outbreak, and they constantly forced themselves through this crust, pouring out melted materials that subsequently cooled and consolidated on its surface. So constant were these eruptions, and so slight was the resistance they encountered, that some portions of the earlier rock-deposits are perforated with numerous chimneys, narrow tunnels as it were, bored by the liquid masses that poured out through them and greatly modified their first condition.

The question at once suggests itself, How was even this thin crust formed? what should cause any solid envelope, however slight and filmy when compared to the whole bulk of the globe, to form upon the surface of such a liquid mass? At this point of the investigation the geologist must appeal to the astronomer; for in this vague and nebulous border-land, where the very rocks lose their outlines and flow into each other, not yet specialized into definite forms and substances, — there the two sciences meet. Astronomy shows us our planet thrown off from the central mass of which it once formed a part, to move henceforth in an independent orbit of its own. That orbit, it tells us, passed through celestial spaces cold enough to chill this heated globe, and of course to consolidate it externally. We know, from the action of similar causes on a smaller scale and on comparatively insignificant objects immediately about us, what must have been the effect of this cooling process upon the heated mass of the globe. All substances when heated occupy more space than they do when cold. Water, which expands when freezing, is the only exception to this rule. The first effect of cooling the surface of our planet must have been to solidify it, and thus to form a film or crust over it. That crust would shrink as the cooling process went on; in consequence of the shrinking, wrinkles and folds would arise upon it, and here and there, where the tension was too great, cracks and fissures would be produced. In proportion as the surface cooled, the masses within would be affected by the change of temperature outside of them, and would consolidate internally also, the crust gradually thickening by this process.

But there was another element without the globe, equally powerful in building it up. Fire and water wrought together in this work, if not always harmoniously, at least with equal force and persistency. I have said that there was a time when no atmosphere surrounded the earth; but one of the first results of the cooling of its crust must have been the formation of an atmosphere, with all the phenomena connected with it, — the rising of vapors, their condensation into clouds, the falling of rains, the gathering of waters upon its surface. Water is a very active agent of destruction, but it works over again the materials it pulls down or wears away, and builds them up anew in other forms. As soon as an ocean washed over the consolidated crust of the globe, it would begin to abrade the surfaces upon which it moved, gradually loosening and detaching materials, to deposit them again as sand or mud or pebbles at its bottom in successive layers, one above another. Thus, in analyzing the crust of the globe, we find at once two kinds of rocks, the respective work of fire and water: the first poured out from the furnaces within, and cooling, as one may see any mass of metal cool that is poured out from a smelting-furnace to-day, in solid crystalline masses, without any division into separate layers or leaves; and the latter in successive beds, one over another, the heavier materials below, the lighter above, or sometimes in alternate layers, as special causes may have determined successive deposits of lighter or heavier materials at some given spot.

There were many well-fought battles between geologists before it was understood that these two elements had been equally active in building up the crust of the earth. The ground was hotly contested by the disciples of the two geological schools, one of which held that the solid envelope of the earth was exclusively due to the influence of fire, while the other insisted that it had been accumulated wholly under the agency of water. This difference of opinion grew up very naturally; for the great leaders of the two schools lived in different localities, and pursued their investigations over regions where the geological phenomena were of an entirely opposite character, — the one exhibiting the effect of volcanic eruptions, the other that of stratified deposits. It was the old story of the two knights on opposite sides of the shield, one swearing that it was made of gold, the other that it was made of silver, and almost killing each other before they discovered that it was made of both. So prone are men to hug their theories and shut their eyes to any antagonistic facts, that it is related of Werner, the great leader of the Aqueous school, that he was actually on his way to see a geological locality of especial interest, but, being told that it confirmed the views of his opponents, he turned round and went home again, refusing to see what might force him to change his opinions. If the rocks did not confirm his theory, so much the worse for the rocks, — he would none of them. At last it was found that the two great chemists, fire and water, had worked together in the vast laboratory of the globe, and since then scientific men had decided to work together also; and if they still have a passage at arms occasionally over some doubtful point, yet the results of their investigations are ever drawing them nearer to each other, — since men who study truth, when they reach their goal, must always meet at last on common ground.

The rocks formed under the influence of heat are called, in geological language, the Igneous, or, as some naturalists have named them, the Plutonic rocks, alluding to their fiery origin, while the others have been called Aqueous or Neptunic rocks, in reference to their origin under the agency of water. A simpler term, however, quite as distinctive, and more descriptive of their structure, is that of the stratified and massive or unstratified rocks. We shall see hereafter how the relative position of these two classes of rocks and their action upon each other enable us to determine the chronology of the earth, to compare the age of her mountains, and, if we have no standard by which to estimate the positive duration of her continents, to say at least which was the first-born among them, and how their characteristic features have been successively worked out. I am aware that many of these inferences, drawn from what is called "the geological record," must seem to be the work of the imagination. In a certain sense this is true, — for imagination, chastened by correct observation, is our best guide in the study of Nature. We are too apt to associate the exercise of this faculty with works of fiction, while it is in fact the keenest detective of truth.

Besides the stratified and massive rocks, there is still a third set, produced by the contact of these two, and called, in consequence of the changes thus brought about, the Metamorphic rocks. The effect of heat upon clay is to bake it into slate; limestone under the influence of heat becomes quick-lime, or, if subjected afterwards to the action of water, it is changed to mortar; sand under the same agency is changed to a coarse kind of glass. Suppose, then, that a volcanic eruption takes place in a region of the earth's surface where successive layers of limestone, of clay, and of sandstone have been previously deposited by the action of water. If such an eruption has force enough to break through these beds, the hot, melted masses will pour out through the rent, flow over its edges, and fill all the lesser cracks and fissures produced by such a disturbance. What will be the effect upon the stratified rocks? Wherever these liquid masses, melted by a heat more intense than can be produced by any artificial means, have flowed over them or cooled in immediate contact with them, the clays will be changed to slate, the limestone will have assumed a character more like marble, while the sandstone will be vitrified. This is exactly what has been found to be the case, wherever the stratified rocks have been penetrated by the melted masses from beneath. They have been themselves partially melted by the contact, and when they have cooled again, their stratification, though still perceptible, has been partly obliterated, and their substance changed. Such effects may often be traced in dikes, which are only the cracks in rocks filled by materials poured into them at some period of eruption when the melted masses within the earth were thrown out and flowed like water into any inequality or depression of the surface around. The walls enclosing such a dike are often found to be completely altered by contact with its burning contents, and to have assumed a character quite different from the rocks of which they make a part; while the mass itself which fills the fissure shows by the character of its crystallization that it has cooled more quickly on the outside, where it meets the walls, than at the centre.

The first two great classes of rocks, the unstratified and stratified rocks, represent different epochs in the world's physical history: the former mark its revolutions, while the latter chronicle its periods of rest. All mountains and mountain-chains have been upheaved by great convulsions of the globe, which rent asunder the surface of the earth, destroyed the animals and plants living upon it at the time, and were then succeeded by long intervals of repose, when all things returned to their accustomed order, ocean and river deposited fresh beds in uninterrupted succession, the accumulation of materials went on as before, a new set of animals and plants were introduced, and a time of building up and renewing followed the time of destruction. These periods of revolution are naturally more difficult to decipher than the periods of rest; for they have so torn and shattered the beds they uplifted, disturbing them from their natural relations to each other, that it is not easy to reconstruct the parts and give them coherence and completeness again. But within the last half-century this work has been accomplished in many parts of the world with an amazing degree of accuracy, considering the disconnected character of the phenomena to be studied; and I think I shall be able to convince my readers that the modern results of geological investigation are perfectly sound logical inferences from well-established facts. In this, as in so many other things, we are but "children of a larger growth." The world is the geologist's great puzzle-box; he stands before it like the child to whom the separate pieces of his puzzle remain a mystery till he detects their relation and sees where they fit, and then his fragments grow at once into a connected picture beneath his hand.

It is a curious fact in the history of progress, that, by a kind of intuitive insight, the earlier observers seem to have had a wider, more comprehensive recognition of natural phenomena as a whole than their successors, who far excel them in their knowledge of special points, but often lose their grasp of broader relations in the more minute investigation of details. When geologists first turned their attention to the physical history of the earth, they saw at once certain great features which they took to be the skeleton and basis of the whole structure. They saw the great masses of granite forming the mountains and mountain-chains, with the stratified rocks resting against their slopes; and they assumed that granite was the first primary agent, and that all stratified rocks must be of a later formation. Although this involved a partial error, as we shall see hereafter, when we trace the upheavals of granite even into comparatively modern periods, yet it held an important geological truth also; for, though granite formations are by no means limited to those early periods, they are nevertheless very characteristic of them, and are indeed the foundation-stones on which the physical history of the globe is built.

Starting from this landmark, the earlier geologists divided the world's history into three periods. As the historian recognizes as distinct phases in the growth of the human race Ancient History, the Middle Ages, and Modern History, so they distinguish between what they call the Primary period, when, as they believed, no life stirred on the surface of the earth; the Secondary or middle period, when animals and plants were introduced and the land began to assume continental proportions; and the Tertiary period, or comparatively modern geological times, when the aspect of the earth as well as its inhabitants was approaching more nearly to the present condition of things. But as their investigations proceeded, they found that every one of these great ages of the world's history was divided into numerous lesser epochs, each of which had been characterized by a peculiar set of animals and plants, and had been closed by some great physical convulsion, disturbing and displacing the materials accumulated during such a period of rest.

The further study of these subordinate periods showed that what had been called Primary formations, the volcanic or Plutonic rocks, formerly believed to be confined to the first geological ages, belonged to all the periods, successive eruptions having taken place at all times, pouring up through the accumulated deposits, penetrating and injecting their cracks, fissures, and inequalities, as well as throwing out large masses on the surface. Up to our own day there has never been a period when such eruptions have not taken place, though they have been constantly diminishing in frequency and extent. In consequence of this discovery, that rocks of igneous character were by no means exclusively characteristic of the earliest times, they are now classified together upon very different grounds from those on which geologists first united them; though, as the name *Primary* was long retained, we still find it applied to them, even in geological works of quite recent date. This defect of nomenclature is to be regretted, as likely to mislead the student, because it seems to refer to time; whereas it no longer signifies the age of the rocks, but simply their character. The name Plutonic or Massive rocks is, however, now almost universally substituted for that of Primary.

A wide field of investigation still remains to be explored by the chemist and the geologist together, in the mineralogical character of the Plutonic rocks, which differs greatly in the different periods. The earlier eruptions seem to have been chiefly granitic, though this must not be understood in too wide a sense, since there are granite formations even as late as the Tertiary period; those of the middle periods were mostly porphyries and basalts; while in the more recent ones, lavas predominate. We have as yet no clue to the laws by which this distribution of volcanic elements in the formation of the earth is regulated; but there is found to be a difference in the crystals of the Plutonic rocks belonging to different ages, which, when fully understood, may enable us to determine the age of any Plutonic rock by its mode of crystallization; so that the mineralogist will as readily tell you by its crystals whether a bit of stone of igneous origin belongs to this or that period of the world's history, as the palæontologist will tell you by its fossils whether a piece of rock of aqueous origin belongs to the Silurian or Devonian or Carboniferous deposits.

Although subsequent investigations have multiplied so extensively not only the number of geological periods, but also the successive creations that have characterized them, yet the first general division into three great eras was nevertheless founded upon a broad and true generalization. In the first stratified rocks in which any organic remains are found, the highest animals are fishes, and the highest plants are cryptogams; in the middle periods reptiles come in, accompanied by fern and moss forests; in later times quadrupeds are introduced, with a dicotyledonous vegetation. So closely does the march of animal and vegetable life keep pace with the material progress of the world, that we may well consider these three divisions, included under the first general classification of its physical history, as the three Ages of Nature; the more important epochs which subdivide them may be compared to so many great dynasties, while the lesser periods are the separate reigns contained therein. Of such epochs there are ten, well known to geologists; of the lesser periods about sixty are already distinguished, while many more loom up from the dim regions of the past, just discerned by the eye of science, though their history is not yet unravelled.

Before proceeding further, I will enumerate the geological epochs in their succession, confining myself, however, to such as are perfectly well established, without alluding to those of which the limits are less definitely determined, and which are still subject to doubts and discussions among geologists. As I do not propose to make here any treatise of Geology, but simply to place before my readers some pictures of the old world, with the animals and plants that have inhabited it at various times, I shall avoid, as far as possible, all debatable ground, and confine myself to those parts of my subject which are best known, and can therefore be more clearly presented.

First, we have the Azoic period, *devoid of life*, as its name signifies,— namely, the earliest stratified deposits upon the heated film forming the first solid surface of the earth, in which no trace of living thing has ever been found. Next comes the Silurian period, when the crust of the earth had thickened and cooled sufficiently to render the existence of animals and plants upon it possible, and when the atmospheric conditions necessary to their maintenance were already established. Many of the names given to these periods are by no means significant of their character, but are merely the result of accident: as, for instance, that of Silurian, given by Sir Roderick Murchison to this set of beds, because he first studied them in that part of Wales occupied by the ancient tribe of the Silures. The next period, the Devonian, was for a similar reason named after the county of Devonshire, in England, where it was first investigated. Upon this follows the Carboniferous period, with the immense deposits of coal from which it derives its name. Then comes the Permian period, named, again, from local circumstances, the first investigation of its deposits having taken place in the province of Permia in Russia. Next in succession we have the Triassic period, so called from the trio of rocks, the red sandstone, Muschel Kalk (shell-limestone), and Keuper (clay), most frequently combined in its formations; the Jurassic, so amply illustrated in the chain of the Jura, where geologists first found the clue to its history; and the Cretaceous period, to which the chalk cliffs of England and all the extensive chalk deposits belong. Upon these follow the so-called Tertiary formations, divided into three periods, all of which have received most characteristic names. In this epoch of the world's history we see the first approach to a condition of things resembling that now prevailing, and Sir Charles Lyell has most fitly named its three divisions, the Eocene, Miocene, and Pliocene. The termination of the three words is made from the Greek word *Kainos*, recent; while *Eos* signifies dawn, *Meion* less, and *Pleion* more. Thus Eocene indicates the dawn of recent species, Pliocene their increase, while Miocene, the intermediate term, means less recent. Above these deposits comes what has been called in science the present period, — *the modern times* of the geologist, — that period to which man himself belongs, and since the beginning of which, though its duration be counted by hundreds of thousands of years, there has been no alteration in the general configuration of the earth, consequently no important modification of its climatic conditions, and no change in the animals and plants inhabiting it.

I have spoken of the first of these periods, the Azoic, as having been absolutely devoid of life, and I believe this statement to be strictly true; but I ought to add that there is a difference of opinion among geologists upon this point, many believing that the first surface of our globe may have been inhabited by living beings, but that all traces of their existence have been obliterated by the eruptions of melted materials, which not only altered the character of those earliest stratified rocks, but destroyed all the organic remains contained in them. It will be my object to show in this series of papers, not only that the absence of the climatic and atmospheric conditions essential to organic life, as we understand it, must have rendered the previous existence of any living beings impossible, but also that the completeness of the Animal Kingdom in those deposits where we first find organic remains, its intelligible and coherent connection with the successive creations of all geological times and with the animals now living, afford the strongest internal evidence that we have indeed found in the lower Silurian formations, immediately following the Azoic, the beginning of life upon earth. When a story seems to us complete and consistent from the beginning to the end, we shall not seek for a first chapter, even though the copy in which we have read it be so torn and defaced as to suggest the idea that some portion of it may have been lost. The unity of the work, as a whole, is an incontestable proof that we possess it in its original integrity. The validity of this argument will be recognized, perhaps, only by those naturalists to whom the Animal Kingdom has begun to appear as a connected whole. For those who do not see order in Nature it can have no value.

For a table containing the geological periods in their succession, I would refer to any modern text-book of Geology, or to an article in the "Atlantic Monthly" for March, 1862, upon "Methods of Study in Natural History," where they are given in connection with the order of introduction of animals upon earth.

Were these sets of rocks found always in the regular sequence in which I have enumerated them, their relative age would be easily determined, for their superposition would tell the whole story: the lowest would, of course, be the oldest, and we might follow without difficulty the ascending series, till we reached the youngest and uppermost deposits. But their succession has been broken up by frequent and violent alterations in the configuration of the globe. Land and water have changed their level, — islands have been transformed to continents, — sea-bottoms have become dry land, and dry land has sunk to form sea-bottom, — Alps and Himalayas, Pyrenees and Apennines, Alleghanies and Rocky Mountains, have had their stormy birthdays since many of these beds have been piled one above another, and there are but few spots on the earth's surface where any number of them may be found in their original order and natural position. When we remember that Europe, which lies before us on the map as a continent, was once an archipelago of islands, — that, where the Pyrenees raised their rocky barrier between France and Spain, the waters of the Mediterranean and Atlantic met, — that, where the British Channel flows, dry land united England and France, and Nature in those days made one country of the lands parted since by enmities deeper than the waters that run between, — when we remember, in short, all the fearful convulsions that have torn asunder the surface of the earth, as if her rocky record had indeed been written on paper, we shall find a new evidence of the intellectual unity which holds together the whole physical history of the globe in the fact that through all the storms of time the investigator is able to trace one unbroken thread of thought from the beginning to the present hour.

The tree is known by its fruits, — and the fruits of chance are incoherence, incompleteness, unsteadiness, the stammering utterance of blind, unreasoning force. A coherence that binds all the geological ages in one chain, a stability of purpose that completes in the beings born to-day an intention expressed in the first creatures that swam in the Silurian ocean or crept upon its shores, a steadfastness of thought, practically recognized by man, if not acknowledged by him, whenever he traces the intelligent connection between the facts of Nature and combines them into what he is pleased to call his system of Geology, or Zoölogy, or Botany, — these things are not the fruits of chance or of an unreasoning force, but the legitimate results of intellectual power. There is a singular lack of logic, as it seems to me, in the views of the materialistic naturalists. While they consider classification, or, in other words, their expression of the relations between animals or between physical facts of any kind, as the work of their intelligence, they believe the relations themselves to be the work of physical causes. The more direct inference surely is, that, if it requires an intelligent mind to recognize them, it must have required an intelligent mind to establish them. These relations existed before man was created; they have existed ever since the beginning of time; hence, what we call the classification of facts is not the work of his mind in any direct original sense, but the recognition of an intelligent action prior to his own existence.

There is, perhaps, no part of the world, certainly none familiar to science, where the early geological periods can be studied with so much ease and precision as in the United States. Along their northern borders, between Canada and the United States, there runs the low line of hills known as the Laurentian Hills. Insignificant in height, nowhere rising more than fifteen hundred or two thousand feet above the level of the sea, these are nevertheless the first mountains that broke the uniform level of the earth's surface and lifted themselves above the waters. Their low stature, as compared with that of other more lofty mountain-ranges, is in accordance with an invariable rule, by which the relative age of mountains may be estimated. The oldest mountains are the lowest, while the younger and more recent ones tower above their elders, and are usually more torn and dislocated also. This is easily understood, when we remember that all mountains and mountain-chains are the result of upheavals, and that the violence of the outbreak must have been in proportion to the strength of the resistance. When the crust of the earth was so thin that the heated masses within easily broke through it, they were not thrown to so great a height, and formed comparatively low elevations, such as the Canadian hills or the mountains of Bretagne and Wales. But in later times, when young, vigorous giants, such as the Alps, the Himalayas, or, later still, the Rocky Mountains, forced their way out from their fiery prison-house, the crust of the earth was much thicker, and fearful indeed must have been the convulsions which attended their exit.

The Laurentian Hills form, then, a granite range, stretching from Eastern Canada to the

Upper Mississippi, and immediately along its base are gathered the Azoic deposits, the first stratified beds, in which the absence of life need not surprise us, since they were formed beneath a heated ocean. As well might we expect to find the remains of fish or shells or crabs at the bottom of geysers or of boiling springs, as on those early shores bathed by an ocean of which the heat must have been so intense. Although, from the condition in which we find it, this first granite range has evidently never been disturbed by any violent convulsion since its first upheaval, yet there has been a gradual rising of that part of the continent, for the Azoic beds do not lie horizontally along the base of the Laurentian Hills in the position in which they must originally have been deposited, but are lifted and rest against their slopes. They have been more or less dislocated in this process, and are greatly metamorphized by the intense heat to which they must have been exposed. Indeed, all the oldest stratified rocks have been baked by the prolonged action of heat.

It may be asked how the materials for those first stratified deposits were provided. In later times, when an abundant and various soil covered the earth, when every river brought down to the ocean, not only its yearly tribute of mud or clay or lime, but the *débris* of animals and plants that lived and died in its waters or along its banks, when every lake and pond deposited at its bottom in successive layers the lighter or heavier materials floating in its waters and settling gradually beneath them, the process by which stratified materials are collected and gradually harden into rock is more easily understood. But when the solid surface of the earth was only just beginning to form, it would seem that the floating matter in the sea can hardly have been in sufficient quantity to form any extensive deposits. No doubt there was some abrasion even of that first crust; but the more abundant source of the earliest stratification is to be found in the submarine volcanoes that poured their liquid streams into the first ocean. At what rate these materials would be distributed and precipitated in regular strata it is impossible to determine; but that volcanic materials were so deposited in layers is evident from the relative position of the earliest rocks. I have already spoken of the innumerable chimneys perforating the Azoic beds, narrow outlets of Plutonic rock, protruding through the earliest strata. Not only are such funnels filled with the crystalline mass of granite that flowed through them in a liquid state, but it has often poured over their sides, mingling with the stratified beds around. In the present state of our knowledge, we can explain such appearances only by supposing that the heated materials within the earth's crust poured out frequently, meeting little resistance,—that they then scattered and were precipitated in the ocean around, settling in successive strata at its bottom,—that through such strata the heated masses within continued to pour again and again, forming for themselves the chimney-like outlets above mentioned.

Such, then, was the earliest American land,— a long, narrow island, almost continental in its proportions, since it stretched from the eastern borders of Canada nearly to the point where now the base of the Rocky Mountains meets the plain of the Mississippi Valley. We may still walk along its ridge and know that we tread upon the ancient granite that first divided the waters into a northern and southern ocean; and if our imaginations will carry us so far, we may look down toward its base and fancy how the sea washed against this earliest shore of a lifeless world. This is no romance, but the bald, simple truth; for the fact that this granite band was lifted out of the waters so early in the history of the world, and has not since been submerged, has, of course, prevented any subsequent deposits from forming above it. And this is true of all the northern part of the United States. It has been lifted gradually, the beds deposited in one period being subsequently raised, and forming a shore along which those of the succeeding one collected, so that we have their whole sequence before us. In regions where all the geological deposits, Silurian, Devonian, Carboniferous, Permian, Triassic, etc., are piled one upon another, and we can get a glimpse of their internal relations only where some rent has laid them open, or where their ragged edges, worn away by the abrading action of external influences, expose to view their successive layers, it must, of course, be more difficult to follow their connection. For this reason the American continent offers facilities to the geologist denied to him in the so-called Old World, where the earlier deposits are comparatively hidden, and the broken character of the land, intersected by mountains in every direction, renders his investigation still more difficult. Of course, when I speak of the geological deposits as so completely unveiled to us here, I do not forget the sheet of drift which covers the continent from North to South, and which we shall discuss hereafter, when I reach that part of my subject. But the drift is only a superficial and recent addition to the soil, resting loosely above the other geological deposits, and arising, as we shall see, from very different causes.

In this article I have intended to limit myself to a general sketch of the formation of the Laurentian Hills with the Azoic stratified beds rest ing against them. In the Silurian epoch following the Azoic we have the first beach on which any life stirred; it extended along the base of the Azoic beds, widening by its extensive deposits the narrow strip of land already upheaved. I propose in my next article to invite my readers to a stroll with me along that beach.

II.

THE SILURIAN BEACH.

WITH what interest do we look upon any relic of early human history! The monument that tells of a civilization whose hieroglyphic records we cannot even decipher, the slightest trace of a nation that vanished and left no sign of its life except the rough tools and utensils buried in the old site of its towns or villages, arouses our imagination and excites our curiosity. Men gaze with awe at the inscription on an ancient Egyptian or Assyrian stone; they hold with reverential touch the yellow parchment-roll whose dim, defaced characters record the meagre learning of a buried nationality; and the announcement, that for centuries the tropical forests of Central America have hidden within their tangled growth the ruined homes and temples of a past race, stirs the civilized world with a strange, deep wonder.

To me it seems, that to look on the first land that was ever lifted above the waste of waters, to follow the shore where the earliest animals and plants were created when the thought of God first expressed itself in organic forms, to hold in one's hand a bit of stone from an old sea-beach, hardened into rock thousands of centuries ago, and studded with the beings that once crept upon its surface or were stranded there by some retreating wave, is even of deeper interest to men than the relics of their own race, for these things tell more directly of the thoughts and creative acts of God.

Standing in the neighborhood of Whitehall, near Lake George, one may look along such a sea-shore, and see it stretching westward and sloping gently southward as far as the eye can reach. It must have had a very gradual slope, and the waters must have been very shallow; for at that time no great mountains had been uplifted, and deep oceans are always the concomitants of lofty heights. We do not, however, judge of this by inference merely; we have an evidence of the shallowness of the sea in those days in the character of the shells found in the Silurian deposits, which shows that they belonged in shoal waters.

Indeed, the fossil remains of all times tell us almost as much of the physical condition of the world at different epochs as they do of its animal and vegetable population. When Robinson Crusoe first caught sight of the footprint on the sand, he saw in it more than the mere footprint, for it

spoke to him of the presence of men on his desert
island. We walk on the old geological shores,
like Crusoe along his beach, and the footprints
we find there tell us, too, more than we actually
see in them. The crust of our earth is a great
cemetery, where the rocks are tombstones on
which the buried dead have written their own
epitaphs. They tell us not only who they were
and when and where they lived, but much also of
the circumstances under which they lived. We
ascertain the prevalence of certain physical conditions at special epochs by the presence of animals
and plants whose existence and maintenance required such a state of things, more than by any
positive knowledge respecting it. Where we find
the remains of quadrupeds corresponding to our
ruminating animals, we infer not only land, but
grassy meadows also, and an extensive vegetation; where we find none but marine animals,
we know the ocean must have covered the earth;
the remains of large reptiles, representing, though
in gigantic size, the half aquatic, half terrestrial
reptiles of our own period, indicate to us the existence of spreading marshes still soaked by the
retreating waters; while the traces of such animals as live now in sand and shoal waters, or in
mud, speak to us of shelving sandy beaches and
of mud-flats. The eye of the Trilobite tells us
that the sun shone on the old beach where he

lived; for there is nothing in nature without a purpose, and when so complicated an organ was made to receive the light, there must have been light to enter it. The immense vegetable deposits in the Carboniferous period announce the introduction of an extensive terrestrial vegetation; and the impressions left by the wood and leaves of the trees show that these first forests must have grown in a damp soil and a moist atmosphere. In short, all the remains of animals and plants hidden in the rocks have something to tell of the climatic conditions and the general circumstances under which they lived, and the study of fossils is to the naturalist a thermometer by which he reads the variations of temperature in past times, a plummet by which he sounds the depths of the ancient oceans,— a register, in fact, of all the important physical changes the earth has undergone.

But although the animals of the early geological deposits indicate shallow seas by their similarity to our shoal-water animals, it must not be supposed that they are by any means the same. On the contrary, the old shells, crustacea, corals, etc., represent types which have existed in all times with the same essential structural elements, but under different specific forms in the several geological periods. And here it may not be amiss to say something of what are called by naturalists *representative types.*

The statement that different sets of animals and plants have characterized the successive epochs is often understood as indicating a difference of another kind than that which distinguishes animals now living in different parts of the world. This is a mistake. They are so-called representative types all over the globe, united to each other by structural relations and separated by specific differences of the same kind as those that unite and separate animals of different geological periods. Take, for instance, mud-flats or sandy shores in the same latitudes of Europe and America; we find living on each animals of the same structural character and of the same general appearance, but with certain specific differences, as of color, size, external appendages, etc. They represent each other on the two continents. The American wolves, foxes, bears, rabbits, are not the same as the European, but those of one continent are as true to their respective types as those of the other; under a somewhat different aspect they represent the same groups of animals. In certain latitudes, or under conditions of nearer proximity, these differences may be less marked. It is well known that there is a great monotony of type, not only among animals and plants, but in the human races also, throughout the Arctic regions; and some animals characteristic of the high North reappear under such identical forms in the neighborhood of the snow-fields in lofty mountains, that to trace the difference between the ptarmigans, rabbits, and other gnawing animals of the Alps, for instance, and those of the Arctics, is among the most difficult problems of modern science.

And so is it also with the animated world of past ages; in similar deposits of sand, mud, or lime, in adjoining regions of the same geological age, identical remains of animals and plants may be found; while at greater distances, but under similar circumstances, representative species may occur. In very remote regions, however, whether the circumstances be similar or dissimilar, the general aspect of the organic world differs greatly, remoteness in space being thus in some measure an indication of the degree of affinity between different faunæ. In deposits of different geological periods immediately following each other, we sometimes find remains of animals and plants so closely allied to those of earlier or later periods that at first sight the specific differences are hardly discernible. The difficulty of solving these questions, and of appreciating correctly the differences and similarities between such closely allied organisms, explains the antagonistic views of many naturalists respecting the range of existence of animals, during longer or shorter geological periods; and the superficial way in which dis cussions concerning the transition of species are carried on, is mainly owing to an ignorance of the conditions above alluded to. My own personal observation and experience in these matters have led me to the conviction that every geological period has had its own representatives, and that no single species has been repeated in successive ages.

The laws regulating the geographical distribution of animals, and their combination into distinct zoölogical provinces called faunæ, with definite limits, are very imperfectly understood as yet; but so closely are all things linked together from the beginning till to-day that I am convinced we shall never find the clew to their meaning till we carry on our investigations in the past and the present simultaneously. The same principle according to which animal and vegetable life is distributed over the surface of the earth now, prevailed in the earliest geological periods. The geological deposits of all times have had their characteristic faunæ under various zones, their zoölogical provinces presenting special combinations of animal and vegetable life over certain regions, and their representative types reproducing in different countries, but under similar latitudes, the same groups with specific differences.

Of course, the nearer we approach the beginning of organic life, the less marked do we find the differences to be, and for a very obvious reason. The inequalities of the earth's surface, her mountain-barriers protecting whole continents from the Arctic winds, her open plains exposing others to the full force of the polar blasts, her snug valleys and her lofty heights, her table-lands and rolling prairies, her river-systems and her dry deserts, her cold ocean-currents pouring down from the high North on some of her shores, while warm ones from tropical seas carry their softer influence to others, — in short, all the contrasts in the external configuration of the globe, with the physical conditions attendant upon them, are naturally accompanied by a corresponding variety in animal and vegetable life.

But in the Silurian age, when there were no elevations higher than the Canadian hills, when water covered the face of the earth, with the exception of a few isolated portions lifted above the almost universal ocean, how monotonous must have been the conditions of life! And what should we expect to find on those first shores? If we are walking on a sea-beach to-day, we do not look for animals that haunt the forests or roam over the open plains, or for those that live in sheltered valleys or in inland regions or on mountain-heights. We look for Shells, for Mussels and Barnacles, for Crabs, for Shrimps, for Marine

Worms, for Star-Fishes and Sea-Urchins, and we may find here and there a fish stranded on the sand or tangled in the sea-weed. Let us remember, then, that, in the Silurian period, the world, so far as it was raised above the ocean, was a beach, and let us seek there for such creatures as God has made to live on sea-shores, and not belittle the Creative work, or say that He first scattered the seeds of life in meagre or stinted measure, because we do not find air-breathing animals when there was no fitting atmosphere to feed their lungs, insects with no terrestrial plants to live upon, reptiles without marshes, birds without trees, cattle without grass, all things, in short, without the essential conditions for their existence.

What we do find, — and these, as I shall endeavor to show my readers, in such profusion that it would seem as if God, in the joy of creation, had compensated Himself for a less variety of forms in the greater richness of the early types, — is an immense number of beings belonging to the four primary divisions of the Animal Kingdom, but only to those classes whose representatives are marine, whose home then, as now, was either in the sea or along its shores. In other words, the first organic creation expressed in its totality the structural conception since carried out in such wonderful variety of details, and purposely limited then, because the world, which was to be the home of the higher animals, was not yet made ready to receive them.

I am fully aware that the intimate relations between the organic and physical world are interpreted by many as indicating the absence, rather than the presence, of an intelligent Creator. They argue, that the dependence of animals on material laws gives us the clew to their origin as well as to their maintenance. Were this influence as absolute and unvarying as the purely mechanical action of physical circumstances must necessarily be, this inference might have some pretence to logical probability, — though it seems to me unnecessary, under any circumstances, to resort to climatic influences or the action of any physical laws to explain the thoughtful distribution of the organic and inorganic world, so evidently intended to secure for all beings what best suits their nature and their needs. But the truth is, that, while these harmonious relations underlie the whole creation in such a manner as to indicate a great central plan, of which all things are a part, there is at the same time a freedom, an arbitrary element in the mode of carrying it out, which seems to point to the exercise of an individual will; for, side by side with facts, apparently the direct result of physical laws, are other facts, the nature of which shows a complete independence of external influences.

Take, for instance, the similarity above alluded to between the faunæ of the Arctics and that of the Alps, certainly showing a direct relation between climatic conditions and animal and vegetable life. Yet even there, where the shades of specific difference between many animals and plants of the same class are so slight as to baffle the keenest investigators, we have representative types both in the Animal and Vegetable Kingdoms as distinct and peculiar as those of widely removed and strongly contrasted climatic conditions. Shall we attribute the similarities and the differences alike to physical causes? Compare, for example, the Reindeer of the Arctics with the Ibex and the Chamois, representing the same group in the Alps. Even on mountain-heights of similar altitudes, where not only climate, but other physical conditions would suggest a recurrence of identical animals, we do not find the same, but representative types. The Ibex of the Alps differs, for instance, from that of the Pyrenees, that of the Pyrenees from those of the Caucasus and Himalayas, these again from each other and from that of the Altai.

But perhaps the most conclusive proof that we must seek for the origin of organic life outside of physical causes consists in the permanence of the fundamental types, while the species representing these types have differed in every geological period. Now what we call typical features of structure are in themselves no more stable or permanent than specific features. If physical causes, such as light, heat, moisture, food, habits of life, etc., acting upon individuals, have gradually in successive generations changed the character of the species to which they belong, why not that of the class and the branch also? If we judge this question from the material side at all, we must, in order to judge it fairly, look at it wholly from that point of view. If these specific changes are brought about in this way, it is because external causes have positive permanent effects upon the substances of which animals are built: they have power to change their hair, to change their skin, to change certain external appendages or ornamentations, and any other of those ultimate features which naturalists call specific characters. Now I would ask what there is in the substances out of which class characters are built that would make them less susceptible to such external influences than these specific characters. In many instances the former are more delicate, more sensitive, far more fragile and transient in their material nature than the latter. And yet never, in all the chances and changes of time, have we seen any alteration in the mode of respiration, of reproduction, of circulation, or in any of the systems of organs which characterize the more comprehensive groups of the Animal Kingdom, although they are quite as much under the immediate influence of physical causes as those structural features which have been constantly changing.

The woody fibre of the Pine-trees has had the same structure from the Carboniferous age to this day, while their mode of branching and the forms of their cones and leaves have been different in each period according to their respective species. The combination of rings, the structure of the wings, and the articulations of the legs are the same in the Cockroaches of the Carboniferous age as in those which infest our ships and our dwellings to-day, while the proportion of their parts is on quite another scale. The tissue of the Corals in the Silurian age is identical in chemical combination and organic structure with that of the Corals of our modern reefs, and yet the extensive researches upon this class, for which we are indebted to Milne Edwards and Haime, have not revealed a single species extending through successive geological ages, but show us, on the contrary, that every age has had its own kinds, differing among themselves in the same way as those of the Gulf of Mexico differ now from those of the Indian Ocean and the Pacific.

The scales of the oldest known fishes in the Silurian beds have the same microscopic structure as those of their representative types to-day, and yet I have never seen a single fossil fish presenting the same specific characters in the successive geological epochs. The teeth of the oldest Sharks show the same microscopic structure as those of the present time, and we do not lack opportunities for comparison, since the former are as common in the mountain-limestone of Ireland as are those of the living Sharks on any beach where our fishermen boil them for the sake of their oil, and yet the Sharks appear under different generic and specific forms in each geological age.

But without multiplying examples, which might be adduced, *ad infinitum*, to show permanence of type combined with repeated changes of species, suffice it to say, that, while the general features in the framework of the organic world and the materials of which that framework is built, though quite as subject to the influence of physical external circumstances as any so-called specific features, have remained perfectly intact from the beginning of Creation till now, so that not the smallest difference is to be discerned in these respects between the oldest representatives of the oldest types in the oldest Silurian rocks and their successors through all the geological ages up to the present day, the species have been different in each epoch. And those still deeper ideal relations, the plans or structural conceptions upon which animals are based, are adhered to through all time with a tenacity in strange contrast to the perishableness of the material forms through which they are expressed.

It is surely a fair question to ask the advocates of the transmutation theory, whether they attribute to physical laws the discernment that would lead them to change the specific features, but to respect all those characters by which the higher structural combinations of the Animal Kingdom are preserved without alteration, — in other words, to maintain the organic plan, while constantly diversifying the mode of expressing it. If so, it would perhaps be as well to call such laws by another name, since they show all the comprehensive wisdom of an intelligent Creator. Until they can tell us why certain features of animals and plants are permanent under conditions which, according to their view, have power to change certain other features no more perishable or transient in themselves, the supporters of the development theory will have failed to substantiate their peculiar scientific doctrine.

But this discussion has led us far away from our starting-point, and interrupted our walk along the Silurian beach; let us return to gather a few specimens there, and compare them with the more familiar ones of our own shores. I have said that the beach was a shelving one, and covered of course with shoal waters; but as I have no desire to mislead my readers, or to present truths as generally accepted which are still subject to dispute, I would state here that the parallel ridges trending east to west across the State of New York, considered by some geologists as the successive shores of a receding ocean, are believed by others to be the inequalities on the bottom of a shallow sea. Not only, however, does the general character of these successive terraces suggest the idea that they must have been shores, but the ripple-marks upon them are as distinct as upon any modern beach. The regular rise and fall of the water is registered there in waving, undulating lines as clearly as on the sand-beaches of Newport or Nahant; and we can see on any one of those ancient shores the track left by the waves as they rippled back at ebb of the tide thousands of centuries ago. One can often see where some obstacle interrupted the course of the water, causing it to break around it; and such an indentation even retains the soft, muddy, plastic look that we observe on the present beaches, where the resistance made by any pebble or shell to the retreating wave has given it greater force at that point, so that the sand around the spot is soaked and loosened. There is still another sign, equally familiar to those who have watched the action of water on a beach. Where a shore is very shelving and flat, so that the waves do not recede in ripples from it, but in one unbroken sheet, the sand and small pebbles are dragged and form lines which diverge whenever the water meets an obstacle, thus forming sharp angles on the sand. Such marks are as distinct on the oldest Silurian rocks as if they had been made yesterday. Nor are these the only indications of the same fact. There are certain animals living always upon sandy or muddy shores, which require for their well-being that the beach should be left dry a part of the day. These animals, moving about in the sand or mud from which the water has retreated, leave their tracks there; and if, at such a time, the wind is blowing dust over the beach, and the sun is hot enough to bake it upon the impressions so formed, they are left in a kind of mould. Such trails and furrows, made by small Shells or Crustacea, are also found in plenty on the oldest deposits.

Admitting it, then, to be a beach, let us begin with the lowest type of the Animal Kingdom, and see what Radiates are to be found there. There are plenty of Corals, but they are not the same kinds of Corals as those that build up our reefs and islands now. The modern Coral animals are chiefly Polyps, but the prevailing Corals of the Silurian age were Acalephian Hydroids, animals which indeed resemble Polyps in certain external features, and have been mistaken for them, but which are nevertheless Acalephs by their internal structure. In these Corals the body, instead of being divided into chambers by the vertical partitions so characteristic of the Polyps, is divided at regular distances by horizontal floors.  I subjoin a wood-cut of a Silurian Coral, which does not, however, show the peculiar internal structure, but gives some idea of the general appearance of the old Hydroid Corals. We have but one Acalephian Coral now living, the Millepore; and it was by comparing that with these ancient ones that I first detected their relation to the Acalephs. For the true Acalephs or Jelly-Fishes we shall look in vain; but the presence of the Acalephian Corals establishes the existence of the type, and we cannot expect to find those kinds preserved which are wholly destitute of hard parts. I do not attempt any description of the Polyps proper, because the early Corals of that class are comparatively few, and do not present features sufficiently characteristic to attract the notice of the casual observer.

Of the Echinoderms, the class of Radiates represented now by our Star-Fishes and Sea-Urchins, we may gather any quantity, though the old-fashioned forms are very different from the living ones. I have dwelt at such length in a former article *
on the wonderful beauty and variety of the Crinoids, or "Stone Lilies," as they have been called, from their resemblance to flowers, that I will only briefly allude to them here. The subjoined woodcut represents one with a closed cup; but the number of their different patterns is hardly to be counted, and I would invite any one who questions the abundant expression of life in those days to look at some slabs of ancient limestone in the Zoölogical Museum at Cambridge, where the stems of the Crinoids are tangled together as thickly as sea-weed on the shore. Indeed, some of our rock-deposits consist chiefly of the fragments of their remains.

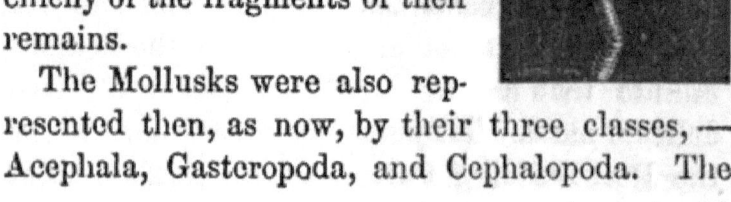

The Mollusks were also represented then, as now, by their three classes, — Acephala, Gasteropoda, and Cephalopoda. The

---

* See Methods of Study in Natural History, Atlantic Monthly, No. LVII., July, 1862.

Acephala or Bivalves we shall find in great numbers, but of a very different pattern from the Oysters, Clams, and Mussels of recent times. The annexed wood-cut represents one of these

Brachiopods, which form a very characteristic type of the Silurian deposits. The square cut of the upper edge, where the two valves meet along the back and are united by a hinge, is altogether old-fashioned, and unknown among our modern Bivalves. The wood-cut does not show the inequality of the two valves, also a very characteristic feature of this group,— one valve being flat and fitting closely into the other, which is more spreading and much fuller. These, also, were represented by a great variety of species, and we find them crowded together as closely in the ancient rocks as Oysters or Clams or Mussels on any of our modern shores. Besides these, there were the Bryozoa, a small kind of compound Mollusk allied to the Clams, and very busy then in the ancient Coral work. They grew in communities, and the separate individuals are so minute that a Bryozoan stock looks like some delicate moss. They still have their place among the Reef-Building Corals, but play an insignificant part in comparison with that of their predecessors.

Of the Silurian Univalves or Gasteropods there is not much to tell, for their spiral shells were so brittle that scarcely any perfect specimens are known, though their broken remains are found in such quantities as to show that this class also was very fully represented in the earliest creation. But the highest class of Mollusks, the Cephalopods or Chambered Shells, or Cuttle-Fishes, as they are called when the animal is unprotected by a shell, are, on the contrary, very well preserved, and they are very numerous. Of these I will speak somewhat more in detail, because their geological history is a very curious one.

The Chambered Nautilus is familiar to all, since, from the exquisite beauty of its shell, it is especially sought for by conchologists; but it is nevertheless not so common in our days as the Squids and Cuttle-Fishes, which are the most numerous modern representatives of the class. In the earliest geological days, on the contrary, those with a shell predominated, differing from the later ones, however, in having the shell perfectly straight instead of curved, though its internal structure was the same as it is now and has ever been. Then, as now, the animal shut himself out from his last year's home, building his annual wall behind him, till his whole shell was divided into successive chambers, all of which were connected by a siphon. Some of the shells of this kind belonging to the Silurian deposits are enormous: giants of the sea they must have been in those days. They have been found fifteen feet long, and as large round as a man's body. One can imagine that the Cuttle-Fish inhabiting such a shell must have been a formidable animal. These straight chambered shells

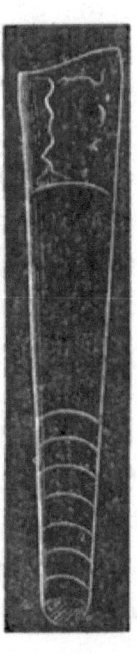

of the Silurian and Devonian seas are called Orthoceratites (see accompanying wood-cut). We shall meet them again hereafter, under another name and with a different form; for, as they advance in the geological ages, they not only assume the curved outline with ever closer whorls till it culminates in the compact coil of the Ammonites of the middle periods, but the partitions, which are perfectly plain walls in these earlier forms, become scalloped and involuted along the edges in the later ones, making the most delicate and exquisite tracery on the surface of the shell.

Of Articulates we find only two classes, Worms and Crustacea. Insects there were none,— for, as we have seen, this early world was wholly marine. There is little to be said of the Worms, for their soft bodies, unprotected by any hard covering, could hardly be preserved; but, like the marine Worms of our own times, they were in the habit of constructing envelopes for themselves, built of sand, or sometimes from a secretion of their own bodies, and these cases we find in the earliest deposits, giving us assurance that the Worms were represented there. I should add, however, that many impressions described as produced by Worms are more likely to have been the tracks of Crustacea.

But by far the most characteristic class of Articulates in ancient times were the Crustaceans. The Trilobites stand in the same relation to the modern Crustacea as the Crinoids do to the modern Echinoderms. They were then the sole representatives of the class, and the variety and richness of the type are most extraordinary. They were of nearly equal breadth for the whole length of the body, and rounded at the two ends, so as to form an oval outline. To give any adequate idea of the number and variety of species would fill a volume, but I may enumerate some of the more striking differences: as, for instance, the greater or less prominence of the anterior shield, — the preponderance of the posterior end in some, while in others the two ends are nearly equal, — the presence or absence of prongs on the shield and of spines along the sides of the body, — appendages on the head in some species, of which others are entirely destitute, — and the smooth outline of some, while in others the surface is broken by a variety of external ornamentation. Such are a few of the more prominent differences among them. But the general structural features are the same in all. The middle region of the body is always divided in uniform rings, lobed in the middle so as to make a ridge along the back with a slight depression on either side of it. It is from this three-lobed division that they receive their name. The subjoined wood-cut represents a characteristic Silurian Trilobite.

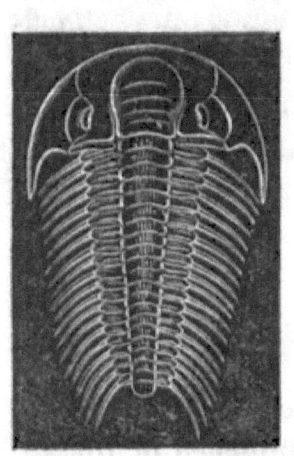

There is no group more prominent in the earliest creations than this one of the Trilobites, and so exclusively do they belong to them, that, as we shall see, in proportion as the later representatives of the class come in, these old-world Crustaceans drop out of the ranks, fall behind, as it were, in the long procession of animals, and are left in the ancient deposits. Even in the Carboniferous period but few are to be found: they had their day in the Silurian and Devonian ages. In consequence of their solid exterior, the preservation of these animals is very complete; and their attitudes are often so natural, and the condition of all their parts so perfect, that one would say they had died yesterday rather than countless centuries ago.

Their geological history has been very thoroughly studied; not only are we familiar with all their adult characters, but even their embryology is well known to naturalists. It is, indeed, wonderful that the mode of growth of animals which died out in the Carboniferous period should be better known to us than that of many living types. But it is nevertheless true that their embryonic forms have been found perfectly preserved in the rocks, and Barrande, in his "Système Silurien de la Bohème," gives us all the stages of their development, from the time when the animal is merely sketched out as a simple furrow in the embryo to its mature condition. So complete is the sequence, that the plate on which their embryonic changes are illustrated contains more than thirty figures, all representing different phases of their growth. There is not a living Crab represented so fully in any of our scientific works as is that one species of Trilobite whose whole story Barrande has traced from the egg to its adult size. Such facts should make those who rest their fanciful theories of the origin and development of life on the imperfection of the geological record, filling up the supposed lapses to suit themselves, more cautious as to their results.

We have found, then, Radiates, Mollusks, and Articulates in plenty; and now what is to be said of Vertebrates in these old times,—of the highest and most important division of the Animal Kingdom, that to which we ourselves belong? They were represented by Fishes alone; and the Fish chapter in the history of the early organic world is a curious, and, as it seems to me, a very significant one. We shall find no perfect specimens; and he would be a daring, not to say a presumptuous thinker, who would venture to reconstruct a fish of the Silurian age from any remains that are left to us. But still we find enough to indicate clearly the style of those old fishes, and to show, by comparison with the living types, to what group of modern times they belong. We should naturally expect to find the Vertebrates introduced in their simplest form; but this is by no means the case: the common fishes, as Cod, Herring, Mackerel, and the like, were unknown in those days.

But there are two groups of so-called fishes, differing from these by some marked features, among which we may find the modern representatives of these earliest Vertebrates. Of these two groups one consists chiefly now of the Gar-

Pikes of our Western waters, though the Sturgeons share also in some of their features. In these fishes there is a singular union of reptilian with fish-like characters. The systems of circulation and of respiration in them are more complicated than in the common fishes; the structure of the skull resembles that of the skull in reptiles, and they have other reptilian characters, such as their ability to move the head upon the neck independently of the body, and the connection of the vertebræ by ball-and-socket joint, instead of by inverted cones, as in the ordinary fishes. Their scales are also peculiar, being covered by enamel so hard, that, if struck with steel, they will emit sparks like flint. It is on account of this peculiarity that the whole group has been called Ganoid. Now, though we have not found as yet any complete specimens of Silurian fishes, their disconnected remains are scattered profusely in the early deposits. The scales, parts of the backbone, parts of the skull, the teeth, are found in a tolerable state of preservation; and these indications, fragmentary as they are, give us the clew to the character of the most ancient fishes. A large proportion of them were no doubt Ganoids; for they had the same peculiar articulation of the vertebræ, the flexibility of the neck, and the hard scales so characteristic of our Gar-Pikes.

There is another type of these ancient Vertebrates which has also its representatives among our modern fishes. These are the Sharks and Skates, or, as the Greeks used to call them, the Selachians,—making a very appropriate distinction between them and common fishes, on account of the difference in the structure of the skeleton. In Selachians the quality of the bones is granular, instead of fibrous, as in fishes; the arches above and below the backbone are formed by flat plates, instead of the spines so characteristic of all the fish proper; and the skull consists of a solid box, instead of being built of overlapping pieces like the true fish-skull. They differ also in their teeth, which, instead of being implanted in the bone by a root, as in fishes, are loosely set in the gum without any connection with the bone, and are movable, being arranged in several rows one behind another, the back rows moving forward to take the place of the front ones when the latter are worn off. They are unlike the common fishes also in having the backbone continued to the very end of the tail, which is cut in uneven lobes, the upper lobe being the longer of the two, while the terminal fin, so constant a feature in fishes, is wanting. The Selachians resemble higher Vertebrate types not only in the small number of their eggs, and in the closer connection of the young with the mother, but also in their embryological development, which has many features in common with that of birds and turtles. Of this group, also, we find numerous remains in the ancient geological deposits; and though we have not the means of distinguishing the species, we have ample evidence for determining the type.

This combination of higher with lower features in the earlier organic forms is very striking, and becomes still more significant when we find that many of the later types recall the more ancient ones. I have called these more comprehensive groups of former times, combining characters of different classes, synthetic or prophetic types; and we might as fitly give the name of retrospective types to many of the later groups, for they recall the past, as the former anticipate the future. And it is not only among the Fishes and the Reptiles that we find these combinations. The most numerous of the ancient Radiates are the Acalephian Corals, combining, in the Hydroid form, the Polyp-like mode of life, habits, and general appearance with the structure of Acalephs. The Crinoids, with the closed cups in some, and the open, star-like crowns in others, unite features of the present Star-Fishes and Sea-Urchins, and, by their stem attaching them to the ground, include also a Polyp-like character; while the Trilobites, with their uniform rings and their prominent anterior shield, unite characters of Worms and Crustacea.

These early types seem to sketch in broad, general characters the Creative purpose, and to include in the first average expression of the plan all its structural possibilities. The Crinoid forms include the thought of the modern Star-Fishes and Sea-Urchins; the simple chambered shells of the Silurian anticipate the more complicated structure of the later ones; the Trilobites give the most comprehensive expression of the Articulate type; while the early Fishes not only prophesy the Reptiles which are to come, but also hint at Birds and even at Mammalia by their embryonic development and their mode of reproduction.

Looked at from this point of view, the animal world is an intellectual Creation, complete in all its parts, and coherent throughout; and when we find, that, although these ancient types have become obsolete and been replaced by modern ones, yet there are always a few old-fashioned individuals, left behind, as it were, to give the key to the history of their race, as the Gar-Pike, for instance, to explain the ancient Fishes, the Millepore to explain the old Acalephian Corals, the Nautilus to be the modern exponent of the Ammonites and Orthoceratites of past times, we cannot avoid the impression that this Creative work has been intended also to be educational for Man, and to teach him his own relation to the organic world. The embryology of the modern types confirms this idea, for here we find an epitome of their geological history. The embryo of the present Star-Fishes recalls the Crinoids; the embryo of the Crab recalls the Trilobites; the embryo of the Vertebrates, including even that of the higher Mammalia, recalls the ancient Fishes. Does not this fact, that the individual animal in its growth recalls the history of its type, prove that the Creative Thought in its immediate present action embraces all that has gone before, as its first organic expression included all that was to come? The study of Nature in its highest meaning shows us the present doubly rich with all the past, and the past linked and interwoven with the present, not lying divorced and dead behind it.

I have spoken of the Silurian beach as if there were but one, not only because I wished to limit my sketch, and to attempt at least to give it the vividness of a special locality, but also because a single such shore will give us as good an idea of the characteristic fauna of the time as if we drew our material from a wider range. There are, however, a great number of parallel ridges belonging to the Silurian and Devonian periods, running from east to west, not only through the

State of New York, but far beyond, through the States of Michigan and Wisconsin into Minnesota; one may follow nine or ten such successive shores in unbroken lines, from the neighborhood of Lake Champlain to the Far West. They have all the irregularities of modern sea-shores, running up to form little bays here, and jutting out in promontories there; and upon each one are found animals of the same kind, but differing in species from those of the preceding.

Although the early geological periods are more legible in North America, because they are exposed over such extensive tracts of land, yet they have been studied in many other parts of the globe. In Norway, in Germany, in France, in Russia, in Siberia, in Kamtschatka, in parts of South America, in short, wherever the civilization of the white race has extended, Silurian deposits have been observed, and everywhere they bear the same testimony to a profuse and varied creation. The earth was teeming then with life as now, and in whatever corner of its surface the geologist finds the old strata, they hold a dead fauna as numerous as that which lives and moves above it. Nor do we find that there was any gradual increase or decrease of any organic forms at the beginning and close of the successive periods. On the contrary, the opening scenes of every chapter in the world's history have been crowded with life, and its last leaves as full and varied as its first.

I think the impression that the faunæ of the early geological periods were more scanty than those of later times arises partly from the fact that the present creation is made a standard of comparison for all preceding creations. Of course, the collections of living types in any museum must be more numerous than those of fossil forms, for the simple reason that almost the whole of the present surface of the earth, with the animals and plants inhabiting it, is known to us, whereas the deposits of the Silurian and Devonian periods are exposed to view only over comparatively limited tracts and in disconnected regions. But let us compare a given extent of Silurian or Devonian sea-shore with an equal extent of sea-shore belonging to our own time, and we shall soon be convinced that the one is as populous as the other. On the New-England coast there are about one hundred and fifty different kinds of fishes, in the Gulf of Mexico two hundred and fifty, in the Red Sea about the same. We may allow in present times an average of two hundred or two hundred and fifty different kinds of fishes to an extent of ocean covering about four hundred miles. Now I have made a special study of the Devonian rocks of Northern Europe, in the Baltic and along the shore of the German Ocean. I have found in those deposits alone one hundred and ten kinds of fossil fishes. To judge of the total number of species belonging to those early ages by the number known to exist now is about as reasonable as to infer that because Aristotle, familiar only with the waters of Greece, recorded less than three hundred kinds of fishes in his limited fishing-ground, therefore these were all the fishes then living. The fishing-ground of the geologist in the Silurian and Devonian periods is even more circumscribed than his, and belongs, besides, not to a living, but to a dead world, far more difficult to decipher.

But the sciences of Geology and Palæontology are making such rapid progress, now that they go hand in hand, that our familiarity with past creations is daily increasing. We know already that extinct animals exist all over the world: heaped together under the snows of Siberia, — lying thick beneath the Indian soil, — found wherever English settlers till the ground or work the mines of Australia, — figured in the old Encyclopædias of China, where the Chinese philosophers have drawn them with the accuracy of their nation, — built into the most beautiful temples of classic lands, for even the stones of the Parthenon are full of the fragments of these old fossils, and if any chance had directed the attention of Aristotle towards them, the science of Palæontology would not have waited for its founder till Cuvier was born,—in short, in every corner of the earth where the investigations of civilized men have penetrated, from the Arctic to Patagonia and the Cape of Good Hope, these relics tell us of successive populations lying far behind our own, and belonging to distinct periods of the world's history.

III.

THE FERN FORESTS OF THE CARBONIFEROUS PERIOD.

DRAW two lines on your map, the upper one running from the mouth of the St. Lawrence westward nearly to St. Paul on the Mississippi, and the lower one from the neighborhood of St. John's in Newfoundland, running southwesterly about to the point where the Wisconsin joins the Mississippi, but jutting down to form an extensive peninsula comprising part of the States of Indiana and Illinois, and you include between them all of the United States which existed at the close of the Devonian period. The upper line rests against the granite hills dividing the Silurian and Devonian deposits of the British Possessions to the north from those of the United States to the south, Canada itself consisting, in great part, of the granite ridge.

How far the early deposits extended to the north of the Laurentian Hills, as well as the out line of that portion of the continent in those times, remains still very problematical; but the investigations thus far undertaken in those regions would lead to the supposition that the same granite upheaval which raised Canada stretched northward in a broad, low ridge of land, widening in its upper part and extending to the neighborhood of Bathurst Inlet and King William's Island, while on either side of it, to the east and west, the Silurian and Devonian deposits extended far toward the present outlines of the continent. These fundamental relations of the continents are admirably presented by Professor Guyot in his charming volume entitled, "Earth and Man."

Indeed, our geological surveys, as well as the information otherwise obtained concerning the primitive condition of North America and the gradual accessions it has received in more recent periods, point to a very early circumscription of the area which, in the course of time, was to become the continent we now inhabit, with its modern features. Not only from the geology of America, but from that of Europe also, it would seem that the position of the continents was sketched out very early in the progressive development of the physical constitution of our earth. It is true that in the present state of our knowledge such wide generalizations must be taken with caution, and held in abeyance to the additional facts which future investigations may develop. But thus far the results certainly do not sustain the theories which have lately found favor among geologists, of entire changes in the relative distribution of land and sea and in the connection of continents with one another; on the contrary, it would appear, that, in accordance with the laws of all organic progress, arising from a fixed starting-point and proceeding through regular changes toward a well-defined end, the continents have grown steadily and consistently from the beginning, through successive accessions in a definite direction, to their present form and organic correlations. If, indeed, there is any meaning in the remarkably symmetrical combinations of the double twin continents in the Eastern Hemisphere, so closely soldered in their northern half, as contrasted with the single pair in the Western Hemisphere, isolated in their position, but so strikingly similar in their outlines, they must be the result of a progressive and predetermined growth already hinted at in the relative position and gradual increase of the first lands raised above the level of the ocean.

However this may be, there can be no doubt that we now know with tolerable accuracy the limits of the land raised above the water in the earlier geological periods in the present United States. Let us see, then, what we enclose between our two lines. We have Newfoundland

and Nova Scotia, the greater part of New England, the whole of New York, a narrow strip along the north of Ohio, a great part of Indiana and Illinois, and nearly the whole of Michigan and Wisconsin.

Within this region lie all the Great Lakes. The origin of these large troughs, holding such immense sheets of fresh water, remains still the subject of discussion and investigation among geologists. It has been supposed that, in the primitive configuration of the globe, when the formation of those depressions at the poles in which the Arctic seas are accumulated gave rise to a corresponding protrusion at the equator, the curve thus produced throughout the North Temperate Zone may have forced up the Canada granite, and have caused, at the same time, those rents in the earth's surface now filled by the Canada lakes; and this view is sustained by the fact that there is a belt of lakes, among which, however, the Canada lakes are far the largest, all around the world in that latitude. The geological phenomena connected with all these lakes have not, however, been investigated with sufficient accuracy and detail, nor has there been any comparison of them extensive and comprehensive enough to justify the adoption of any theory respecting their origin. In an excursion to Lake Superior, some years since, I satisfied myself that the position and outline of that particular lake had their immediate cause in several distinct systems of dikes which intersect its northern shore, and have probably cut up the whole tract of rock over the space now filled by that wonderful sheet of fresh water in such a way as to destroy its continuity, to produce depressions, and gradually create the excavation which now forms the basin of the lake. How far the same causes have been effectual in producing the other large lakes I am unable to say, never having had the opportunity of studying their formation with the same care.

The existence of the numerous smaller lakes running north and south in the State of New York, as the Canandaigua, Seneca, Cayuga, etc., is more easily accounted for. Slow and gradual as was the process by which all that region was lifted above the ocean, it was, nevertheless, accompanied by powerful dislocations of the stratified deposits, as we shall see when we examine them with reference to the local phenomena connected with them. To these dislocations of the strata we owe the transverse cracks across the central part of New York, which needed only the addition of the fresh water poured into them by the rains to transform them into lakes.

I shall not attempt any account of the differences between the animals of the Devonian period and those of the Silurian period, because they consist of structural details difficult to present in a popular form and uninteresting to all but the professional naturalist. Suffice it to say, that, though the organic world had the same general character in these two closely allied periods, yet its representatives in each were specifically distinct, and their differences, however slight, are as constant and as definitely marked as those between more widely separated creations.

At the close of the Devonian period, several upheavals occurred of great significance for the future history of America. One in Ohio raised the elevated ground on which Cincinnati now stands; another hill lifted its granite crest in Missouri, raising with it an extensive tract of Silurian and Devonian deposits; while a smaller one, which does not seem, however, to have disturbed the beds about it so powerfully, broke through in Arkansas. At the same time, elevations took place toward the East, — the first links, few and detached, in the great Alleghany chain which now raises its rocky wall from New England to Alabama.

In the Ohio hill, the granite did not break through, though the force of the upheaval was such as to rend asunder the Devonian deposits, for we find them lying torn and broken about the base of the hill; while the Silurian beds, which should underlie them in their natural position, form its centre and summit. This accounts for the great profusion of Silurian organic remains in that neighborhood. Indeed, there is no locality which forces upon the observer more strongly the conviction of the profusion and richness of the early creation. One may actually collect the remains of Silurian Shells and Crustacea by cart-loads around the city of Cincinnati. A naturalist would find it difficult to gather, along any modern sea-shore, even on tropical coasts, where marine life is more abundant than elsewhere, so rich a harvest, in the same time, as he will bring home from an hour's ramble in the environs of that city.

These elevations naturally gave rise to depressions between themselves and the land on either side of them, and caused also so many counter-slopes dipping toward the uniform southern slope already formed at the north. Thus between the several new upheavals, as well as between them all and the land to the north of them, wide basins or troughs were formed, enclosed on the south, west, and east by low hills, (for these more recent eruptions were, like all the early upheavals, insignificant in height,) and bounded on the north by the more ancient shores of the preceding ages.

These were the inland seas of the Carboniferous period. Here, again, we must infer the successive stages of a history which we can read only in its results. Shut out from the ocean, these shallow sea-basins were gradually changed by the rains to fresh-water lakes; the lakes, in their turn, underwent a transformation, becoming filled, in the course of centuries, with the materials worn away from their shores, with the *débris* of the animals which lived and died in their waters, as well as with the decaying matter from aquatic plants, till at last they were changed to spreading marshes, and on these marshes arose the gigantic fern-vegetation of which the first forests chiefly consisted. Such are the separate chapters in the history of the coal-basins of Illinois, Missouri, Pennsylvania, New England, and Nova Scotia. First inland seas, then fresh-water lakes, then spreading marshes, then gigantic forests, and lastly vast storehouses of coal for the human race.

Although coal-beds are by no means peculiar to the Carboniferous period, since such deposits must be formed wherever the decay of vegetation is going on extensively, yet it would seem that coal-making was the great work in that age of the world's physical history. The atmospheric conditions, so far as we can understand them, were then especially favorable to this result. Though the existence of such an extensive terrestrial vegetation shows conclusively that an atmosphere must have been already established, with all the attendant phenomena of light, heat, air, moisture, etc., yet it is probable that this atmosphere differed from ours in being very largely charged with carbonic acid.

We should infer this from the nature of the animals characteristic of the period; for, though land-animals were introduced, and the organic world was no longer exclusively marine, there were as yet none of the higher beings in whom respiration is an active process. In all warm-blooded animals the breathing is quick, requiring a large proportion of oxygen in the surrounding air, and indicating by its rapidity the animation of the whole system; while the slow-breathing, cold-blooded animals can live in an air that is heavily loaded with carbon. It is well known, however, that, though carbon is so deadly to higher animal life, plants require it in great quantities; and it would seem that one of the chief offices of the early forests was to purify the atmosphere of its undue proportion of carbonic acid, by absorbing the carbon into their own substance, and eventually depositing it as coal in the soil.

Another very important agent in the process of purifying the atmosphere, and adapting it to the maintenance of a higher organic life, is found in the deposits of lime. My readers will excuse me, if I introduce here a very elementary chemical fact to explain this statement. Limestone is carbonate of calcium. Calcium is a metal, fusible as such, and, forming a part of the melted masses within the earth, it was thrown out with the eruptions of Plutonic rocks. Brought to the air, it would appropriate a certain amount of oxygen, and by that process would become oxide of calcium, in which condition it combines very readily with carbonic acid. Thus it becomes carbonate of lime; and all lime deposits played an important part in establishing the atmospheric proportions essential to the existence of the warm-blooded animals.

Such facts remind us how far more comprehensive the results of science will become when the different branches of scientific investigation are pursued in connection with each other. When chemists have brought their knowledge out of their special laboratories into the laboratory of the world, where chemical combinations are and have been through all time going on in such vast proportions, — when physicists study the laws of moisture, of clouds and storms, in past periods as well as in the present, — when, in short, geologists and zoölogists are chemists and physicists, and *vice versâ*, — then we shall learn more of the changes the world has undergone than is possible now that they are separately studied.

It may be asked, how any clew can be found to phenomena so evanescent as those of clouds and moisture. But do we not trace in the old deposits the rain-storms of past times? The heavy drops of a passing shower, the thick, crowded tread of a splashing rain, or the small pin-pricks of a close and fine one, — all the story, in short, of the rising vapors, the gathering clouds, the storms and showers of ancient days, we find recorded for us in the fossil rain-drops; and when we add to this the possibility of analyzing the chemical elements which have been absorbed into the soil, but which once made part of the atmosphere, it is not too much to hope that we shall learn something hereafter of the meteorology even of the earliest geological ages.

The peculiar character of the vegetable tissue in the trees of the Carboniferous period, containing, as it did, a large supply of resin drawn from the surrounding elements, confirms the view of the atmospheric conditions above stated; and this fact, as well as the damp, soggy soil in which the first forests must have grown, accounts for the formation of coal in greater quantity and more combustible in quality than is found in the more recent deposits. But stately as were those fern forests, where plants which creep low at our feet to-day, or are known to us chiefly as underbrush, or as rushes and grasses in swampy grounds, grew to the height of lofty trees, yet the vegetation was of an inferior kind.

There has been a gradation in time for the vegetable as well as the animal world. With the marine population of the more ancient geological ages we find nothing but sea-weeds, — of great variety, it is true, and, as it would seem, from some remains of the marine Cryptogams in early times, of immense size, as compared with modern sea-weeds. But in the Carboniferous period, the plants, though still requiring a soaked and marshy soil, were aërial or atmospheric plants: they were covered with leaves; they breathed; their fructification was like that which now characterizes the ferns, the club-mosses, and the so-called "horsetail plants," (*Equisetaceæ*,) those grasses of low, damp grounds, remarkable for the strongly marked articulations of the stem.

These were the lords of the forests all over the world in the Carboniferous period. Wherever the Carboniferous deposits have been traced, in the United States, in Canada, in England, France, Belgium, Germany, in New Holland, at the Cape of Good Hope, and in South America, the general aspect of the vegetation has been found to be the same, though characterized in the different localities by specific differences of the same nature as those by which the various floræ are distinguished now in different parts of the same zone.

For instance, the Temperate Zone throughout the world is characterized by certain families of trees: by Oaks, Maples, Beeches, Birches, Pines, etc.; but the Oaks, Maples, Beeches, Birches, and the like, of the American flora in that latitude differ in species from the corresponding European flora. So in the Carboniferous period, when more uniform climatic conditions prevailed throughout the world, the character of the vegetation showed a general unity of structure everywhere; but it was nevertheless broken up into distinct botanical provinces by specific differences of the same kind as those which now give such diversity of appearance to the vegetation of the Temperate Zone in Europe as compared with that of America, or to the forests of South America as compared with those of Africa.

There can be no doubt as to the true nature of the Carboniferous forests; for the structural character of the trees is as strongly marked in their fossil remains as in any living plants of the same character. We distinguish the Ferns not only by the peculiar form of their leaves, often perfectly preserved, but also by the fructification on the lower surface of the leaves, and by the distinct marks made on the stem at their point of juncture with it. The leaf of the Fern, when falling, leaves a scar on the stem varying in shape and size according to the kind of Fern, so that the botanist readily distinguishes any particular species of Fern by this means, — a birth-mark, as it were, by which he detects the parentage of the individual. Another indication, equally significant, is found in the tubular structure of the wood in Ferns. On a vertical section of any well-preserved Fern-trunk from the old forests the little tubes may be seen very distinctly running up its length; or, if it be cut through transversely, they may be traced by the little pores like dots on the surface. Trees of this description are found in the Carboniferous marshes, standing erect and perfectly preserved, with trunks a foot and a half in diameter, rising to a height of many feet. Plants so strongly bituminous as the Ferns, when they equalled in size many of our present forest-trees, naturally made coal deposits of the most combustible quality. It is true that we find the anthracite coal of the same period with comparatively little bituminous matter; but this is where the bitumen has been destroyed by the action of the internal heat of the earth.

Next to the Ferns, the Club-Mosses (*Lycopodiaceæ*) seem to have contributed most largely to the marsh-forests. They were characterized, then, as now, by the small size of the leaves growing close against the stem, so that the stem itself, though covered with leaves, looks almost naked, like the stem of the Cactus. Beside these, there are the tree-like Equiseta, in which we find the articulations on the trunk corresponding exactly to those now so characteristic of those marsh-grasses which are the modern representatives of this family of plants, with cone-like fructifications on the summit of the stem.

I would merely touch here upon a subject which does not belong to my own branch of Natural History, but is of the greatest interest in botanical research, namely, the gradation of plants in the geological ages, and the combination of characters in some of the earlier vegetable forms, corresponding to that already noticed in the ancient animal types. For instance, in the Carboniferous period we have only Cryptogams, Ferns, Lycopodiaceæ, and Equisetaceæ. In the middle geological ages, Coniferæ are introduced, the first flowering plant known on earth, but in which the flower and fruit are very imperfect as compared with those of the higher groups. The Coniferæ were chiefly represented in the middle periods by the Cycadæ, that peculiar group of Coniferæ, resembling Pines in their structure, but recalling the Ferns by their external appearance. The stem is round and short, its surface being covered with scars similar to those of the Ferns; while on the summit are ten or more leaves, fan-like and spreading when their growth is complete, but rolled up at first, like Fern-leaves before they expand. Their fruit resembles somewhat the Pine-Apple.

The mode of growth of the Coniferæ recalls a feature of the Equisetaceæ also, in the tufts of little leaves which appear in whorls at regular intervals along the length of the stem in proportion as it elongates, reminding one of the articulations on the stem of the Equisetaceæ. The first cone also appears on the summit of the stem, like the terminal cone in the Equisetaceæ and the Club-Mosses. Thus in certain types of the vegetable, as well as the animal creation of earlier times, there was a combination of features, afterwards divided and presented in separate groups. In the present times, no one of these families of plants overlaps the others, but each has a distinct individual character of its own.

At the close of the middle geological ages and the opening of the Tertiary periods, the Monocotyledons become abundant, the first plants with flower and enclosed seed, though with no true floral envelope; but not until the two last epochs of the Tertiary age do we find in any number the Dicotyledonous plants, in which flower and fruit rise to their highest perfection. Thus there has been a procession of plants, from their earliest introduction to the present day, corresponding to their botanical rank as they now exist, so that the series of gradation in the Vegetable Kingdom, as well as the Animal Kingdom, is the same, whether founded upon succession in time or upon comparative structural rank.

Some attempt has been made to reproduce under an artistic form the aspect of the world in the different geological ages, and to present in single connected pictures the animal and vegetable world of each period. Professor F. Unger, of Vienna, has prepared a collection of fourteen such sketches, entitled, "Tableaux Physionomiques de la Végétation des Diverses Périodes du Monde Primitif."

First, we have the Devonian shores, with spreading fields of sea-weed and numbers of the club-shaped Algæ of gigantic size. He has ventured, also, to represent a few trees, with scanty foliage; but I believe their existence at so early a period to be very problematical.

Next comes the Carboniferous forest, with still pools of water lying between the Fern-trees, which, much as they affect damp, swampy grounds, seem scarcely able to find foothold on the dripping earth. Their trunks, as well as those of the Club-Moss trees which make the foreground of the picture, stand up free from any branches for many feet above the ground, giving one a glimpse between them into the dim recesses of this quiet, watery wood, where the silence was unbroken by the song of birds or the hum of insects. We shall find, it is true, when we give a glance at the animals of this time, that certain insects made their appearance with the first terrestrial vegetation; but they were few in number and of a peculiar kind, such as thrive now in low, wet lands.

Upon this follow a number of sketches introducing us to the middle periods, where the land is higher and more extensive, covered chiefly with Pine-forests, beneath which grows a thick carpet of underbrush, consisting mostly of Grasses, Rushes, and Ferns. Here and there one of the gigantic reptiles of the time may be seen sunning himself on the shore. One of these sketches shows us such a creature hungrily inspecting a pool where Crinoids, with their long stems, large, closely-coiled Chambered Shells, and Brachiopods, the Oysters and Clams of those days, offer him a tempting repast. Here and there a Pterodactyl, the curious winged reptile of the later middle periods, stretches its long neck from the water, and birds also begin to make their appearance.

After these come the Tertiary periods: the Eocene first, where the landscape is already broken up by hills and mountains, clothed with a varied vegetation of comparatively modern character. Lily-pads are floating on the stream which makes the central part of the picture; large herds of the Palæotherium, the ancient Pachyderm, reconstructed with such accuracy by Cuvier, are feeding along its banks; and a tall bird of the Heron or Pelican kind stands watching by the water's edge. In the Miocene the vegetation looks still more familiar, though the Elephants, roaming about in regions of the Temperate Zone, and the huge Salamanders, crawling out of the water, remind us that we are still far removed from present times. Lastly, we have the ice period, with the glaciers coming down to the borders of a river where large troops of Buffalo are drinking, while on the shore some Bears are feasting on the remains of a huge carcass.

It is, however, with the Carboniferous age that we have to do at present, and I will not anticipate the coming chapters of my story by dwelling now on the aspect of the later periods. To return, then, to the period of the coal, it would seem that extensive freshets frequently overflowed the marshes, and that even after many successive forests had sprung up and decayed upon their soil, they were still subject to submergence by heavy floods. These freshets, at certain intervals, are not difficult to understand, when we remember, that, beside the occasional influx of violent rains, the earth was constantly undergoing changes of level, and that a subsidence or upheaval in the neighborhood would disturb the equilibrium of the waters, causing them to overflow and pour over the surface of the country, thus inundating the marshes anew.

That such was the case we can hardly doubt, after the facts revealed by recent investigations of the Carboniferous deposits. In some of the deeper coal-beds there is a regular alternation between layers of coal and layers of sand or clay; in certain localities, as many as ten, twelve, and even fifteen coal-beds have been found alternating with as many deposits of clay or mud or sand; and in some instances, where the trunks of the trees are hollow and have been left standing erect, they are filled to the brim, or to the height of the next layer of deposits, with the materials that have been swept over them. Upon this set of deposits comes a new bed of coal with the remains of a new forest, and above this again a layer of materials left by a second freshet, and so on through a number of alternate strata. It is evident from these facts that there has been a succession of forests, one above another, but that in the intervals of their growth great floods have poured over the marshes, bringing with them all kinds of loose materials, such as sand, pebbles, clay, mud, lime, etc., which, as the freshets subsided, settled down over the coal, filling not only the spaces between such trees as remained standing, but even the hollow trunks of the trees themselves.

Let us give a glance now at the animals which inhabited the waters of this period. In the Radiates we shall not find great changes; the three classes are continued, though with new representatives, and the Polyp Corals are increasing, while the Acalephian Corals, the Rugosa and Tabulata, are diminishing. The Crinoids were still the most prominent representatives of the class of Echinoderms, though some resembling the Ophiurans and Echinoids (Sea-Urchins) began to make their appearance. The adjoining wood-cut represents a characteristic Crinoid of the Carboniferous age.

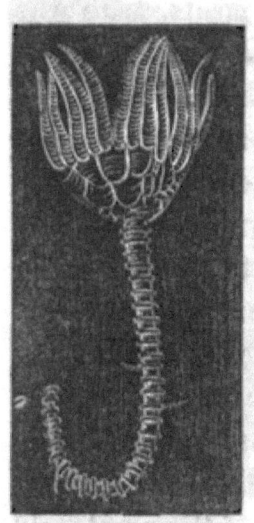

Among the Mollusks, Brachiopods are still prominent, one new genus among them, the Productus, being very remarkable on account of the manner in which one valve rises above the other. The following wood-cut represents such a shell, looked at from the side of the flat valve, showing the straight cut of the line of juncture between the valves and the rising curve of the opposite one, which looks like a hooked beak when seen in profile.

Other species of Bivalves were also introduced, approaching more nearly our Clams and Oysters, or, as they are called in scientific nomenclature, the Lamellibranchiates. They differ  from the Brachiopods chiefly in the higher character of their breathing-apparatus; for they have free gills, instead of the net-work of vessels on the lining skin which serves as the organ of respiration in the Brachiopods. We shall always find, that, in proportion as the functions are distinct, and, as it were, individualized by having special organs appropriated to them, animals rise in the scale of structure. The next class of Mollusks, the Gasteropods, or Univalves, with spiral shells, were numerous, but, from their brittle character, are seldom found in a good state of preservation.

The Chambered Shells, or the Cephalopods, represented chiefly in the earlier periods by the straight Orthoceratites described in a previous article, are now curled in a close coil, and the internal structure of their chambers has become more complicated. The subjoined wood-cut rep-

resents a characteristic Chambered Shell of the Carboniferous age. Goniatites is the scientific name of these later forms. If we had looked for them in the Devonian period, we should have found many with looser coils than these, and some only slightly curved in the shape of a horn. These, as well as the perfectly straight forms, still exist in the coal period, but the Goniatites with close whorls are the more numerous and more characteristic.

The Articulates have gained their missing class since the close of the Devonian period, for Insects have come in, and that division of the Animal Kingdom is therefore complete, and represented by three classes, as it is at present. Of the Worms little can be said; their traces are found as before, but they are very imperfectly preserved. There are still Trilobites, but they are very few in number, and other groups of Crustacea have been added.

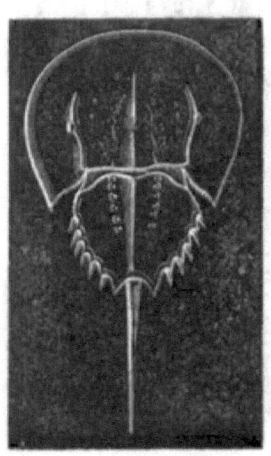

One of the most prominent of these new types bears a striking resemblance to the Horse-Shoe Crab of present times. I here present one of our common Horse-Shoe Crabs above one of these old-world Crustaceans, and it will be seen, that while the latter preserves some of the Trilobitic characters, such as the marked articulations on the posterior part of the body and their division into three lobes, yet, in the prominence of its anterior shield, its more elongated form, and 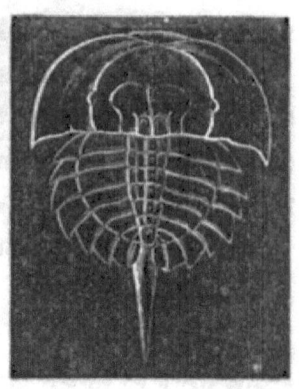 tapering extremity, it resembles its modern representative. In some of them, however, there is no such sharp point as is here figured, and the body terminates bluntly. There were a large number of these Entomostraca in the Carboniferous period, a group which is chiefly represented among living Crustacea by an exceedingly minute kind of Shrimp; but in those days they were of the size of our Crabs and Lobsters, or even larger, and the Horse-Shoe Crab still maintains their claim to a place among the larger and more conspicuous members of the class.

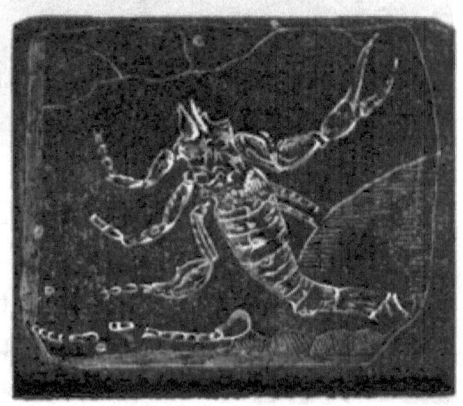

The Insects were few, and, as I have said above, of a kind

which seeks a moist atmosphere, or whose larvæ live altogether in water. They are not usually well preserved, as will be seen from the broken character of the one here represented, although the wood-cut is made from a better specimen than is often found. We have, however, remains enough to establish unquestionably the fact of their existence in the Carboniferous period, and to show us that the type of Articulates was already represented by all its classes.

Not so with the Vertebrates. Fishes abound, but their class still consists, as before, of the Ganoids, those fishes of the earlier periods built on the Gar-Pike and Sturgeon pattern, and the Selachians, represented now by Sharks and Skates. In the Carboniferous period we begin to find perfectly preserved specimens of the Ganoids, and

the adjoining wood-cut represents such a one. Of the old type of Selachians we have again one lingering representative in our own times to give us the clew to its ancestors, — as the Gar-Pike explains the old Ganoids, and the Chambered Nautilus helps us to understand the Chambered Shells of past times. The so-called Port-Jackson

Shark has features which were very characteristic of the Carboniferous Sharks and are lost in the modern ones, so that it affords us a sort of link, as it were, and a measure of comparison, between those now living and the more ancient forms. It is an interesting fact that this only living representative of the Carboniferous Shark should be found in New Holland, because it is there, in that isolated continent, left apart, as it would seem, for a special purpose, that we find reproduced for us most fully the character of the Animal Kingdom in earlier creations.

The first Mammalia in the world were pouched animals, having that extraordinary attachment to the mother after birth which characterizes the Kangaroo. In New Holland almost all the Mammalia are pouched, and have also the imperfect organization of the brain, as compared with the other Mammalia, which accompanies that peculiar structural feature; and although the American Opossum makes an exception to the rule, it is nevertheless true that this type of the Animal Kingdom is now confined almost exclusively to New Holland. Whether this living picture of old creations in modern garb was meant to be educational for man or not, it is at least well that we should take advantage of it in learning all it has to teach us of the relations between the organic world of past and present times.

There were a great variety of the Selachians in the Carboniferous period. The wood-cuts below represent a tooth and a spine from one of the

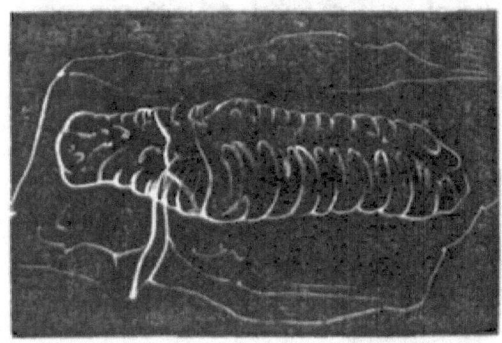

most characteristic groups, but I have not thought it worth while to enumerate or to figure others here, for there are no perfect specimens, and their structural differences consist chiefly in the various form and appearance of the teeth, scales, and spines, and would be uninteresting to most of my readers. I would refer the more scientific ones, who may care to know something of these details, to my investigations on Fossil Fishes, published many years since under the title of "Recherches sur les Poissons Fossiles."

Although the Vertebrate division of the Animal Kingdom still waited for its higher classes, yet it had received one important addition since the Silurian and Devonian periods. The Carboniferous marshes were not without their reptilian inhabitants; but they were Reptiles of the lowest class, the so-called Amphibians, those which are hatched from the egg in an immature condition, undergoing metamorphosis after birth. They have no hard scales, and lay a large number of eggs. I am unable to present any figure of one of these ancient Reptiles, as they are found in so imperfect a state of preservation that no plates have been made from them. I would add, in connection with this subject, that I believe a large number of animals found in the Carboniferous deposits, and referred to the class of Reptiles, to be Fishes allied to Saurians.

Before leaving the Carboniferous period, let us see what territory the United States has conquered from the Ocean during that time. All its central portion, from Canada to Alabama, and from Western Iowa, Missouri, and Arkansas to Eastern Virginia, was raised above the water. But as yet the Alleghanies and the Rocky Mountains did not exist; a great gulf ran up to the mouth of the Ohio, for the Mississippi had not yet accumulated the soil for the fertile valley through which it was to take its southern course; the Coral-Builders had still their work to do in constructing the peninsula of Florida; and, indeed, all the borders of the continent of North

America, as well as a large part of its Western territory, were still to be added. But although its central portion held its ground and was never submerged again, yet the continent was slowly subsiding during the middle geological periods, so that, instead of enlarging gradually by the increase of deposits, its limits remained much the same.

This accounts for the very scanty traces to be found in America of the secondary deposits; for the Permian, Triassic, and Jurassic beds, instead of being raised to form successive shores, along which their deposits could be accumulated in regular sequence, as had been the case with the Azoic, Silurian, and Devonian deposits in the northern part of the United States, were constantly sinking, so that the Triassic settled above the Permian, the Jurassic above the Triassic, and so on, each set of strata thus covering over and concealing the preceding one. Though we find the stratified rocks of these periods cropping out here and there, where some violent disturbance or the abrading action of water has torn asunder or worn away the overlying strata, yet we never find them consecutively over any extensive region; and it is not till the Cretaceous and earlier Tertiary periods that we find again a regular succession of deposits around the shores of the continent, marking its present outlines. It is, then, in Europe, where the sequence of their beds is most complete, that we must seek to decipher the history of the middle geological ages; and therefore, when I meet my readers again, it will be in the Old World of civilization, though more recent in its physical features than the one we leave

IV.

MOUNTAINS AND THEIR ORIGIN.

A CHAPTER on mountains will not be an inappropriate introduction to that part of the world's history on which we are now entering, when the great inequalities of the earth's surface began to make their appearance; and before giving any special account of the geological succession in Europe, I will say something of the formation of mountains in general, and of the men whose investigations first gave us the clew to the intricacies of their structure. It has been the work of the nineteenth century to decipher the history of the mountains, to smooth out these wrinkles in the crust of the earth, to show that there was a time when they did not exist, to decide at least comparatively upon their age, and to detect the forces which have produced them.

But while I speak of the reconstructive labors of the geologist with so much confidence, because to my mind they reveal an intelligible coherence in the whole physical history of the world, yet I am well aware that there are many and wide gaps in our knowledge to be filled up. All the attempts to represent the appearance of the earth in past periods by means of geological maps are, of course, but approximations of the truth, and will compare with those of future times when the phenomena are better understood, much as our present geographical maps, the result of repeated surveys and of the most accurate measurements, compare with those of the ancients.

Homer's world was a flat expanse, surrounded by ocean, of which Greece was the centre. Asia Minor, the Ægean Islands, Egypt, part of Italy and Sicily, the Mediterranean and the Black Sea filled out and completed his map.

Hecatæus, the Greek historian and geographer, who lived more than five hundred years before Christ, had not enlarged it much. He was, to be sure, a voyager on the Mediterranean, and had an idea of the extent of Italy. Acquaintance with Phœnician merchants also had enlarged his knowledge of the world; Sardinia, Corsica, and Spain were known to him, and he was familiar with the Black and Red Seas; though an indentation on his map in the neighborhood of the Caspian would seem to indicate that he was aware of the existence of this sea also, it is not otherwise marked.

Herodotus makes a considerable advance beyond his predecessors: the Caspian Sea has a place on his map; Asia is sketched out, including the Persian Gulf, with the large rivers pouring into it; and the course of the Ganges is traced, though he makes it flow east and empty into the Pacific, instead of turning southward and emptying into the Indian Ocean.

Eratosthenes, two centuries before Christ, is the first geographer who makes some attempt to determine the trend to the land and water, presenting a suggestion that the earth is broader in one direction than in the other. In his map he adds also the geographical results derived from the expeditions of Alexander the Great.

Ptolemy, who flourished in Alexandria in the reign of Hadrian, is the next geographer of eminence, and he shows us something of Africa; for, in his time, the Phœnicians, in their commercial expeditions, had sailed far to the south, had reached the termination of Africa, with ocean lying all around it, and had seen the sun to the north of them. This last assertion, however, Ptolemy does not credit, and he is as sceptical of the open ocean surrounding the extremity of Africa as modern geographers and explorers have been of the existence of Kane's open Arctic Sea. He believes that what the Phœnician traders took to be the broad ocean must be part of an inland sea, corresponding to the Mediteranean, with which he was so familiar. His map includes also England, Ireland, and Scotland; and his Ultima Thule is, no doubt, the Hebrides of our days.

Our present notions of the past periods of the world's history probably bear about the same relation to the truth that these ancient geographical maps bear to the modern ones. But this should not discourage us, for, after all, those maps were in the main true as far as they went; and as the ancient geographers were laying the foundation for all our modern knowledge of the present conformation of the globe, so are the geologists of the nineteenth century preparing the ground for future investigators, whose work will be as far in advance of theirs as are the delineations of Carl Ritter, the great master of physical geography in our age, in advance of the map drawn by the old Alexandrian geographer. We shall have our geological explorers and discoverers in the lands and seas of past times, as we have had in those of the present,— our Columbuses, our Captain Cooks, our Livingstones in geology, as we have had in geography. There are undiscovered continents and rivers and inland seas in the past world to exercise the ingenuity, courage, and perseverance of men, after they shall have solved all the problems, sounded all the depths, and scaled all the heights of the present surface of the earth.

What has been done thus far is chiefly to classify the inequalities of the earth's surface, and to detect the different causes which have produced them. Foldings of the earth's crust, low hills, extensive plains, mountain-chains and narrow valleys, broad table-lands and wide valleys, local chimneys or volcanoes, river-beds, lake-basins, inland seas, — such are some of the phenomena which, disconnected as they seem at first glance, have nevertheless been brought under certain principles, and explained according to definite physical laws.

Formerly men looked upon the earth as a unit in time, as the result of one creative act, with all its outlines established from the beginning. It has been the work of modern science to show that its inequalities are not contemporaneous or simultaneous, but successive, including a law of growth, — that heat and cold, and the consequent expansion and contraction of its crust, have produced wrinkles and folds upon the surface, while constant oscillations, changes of level which are even now going on, have modified its conformation, and moulded its general outline through successive ages.

In thinking of the formation of the globe, we must at once free ourselves from the erroneous impression that the crust of the earth is a solid, steadfast foundation. So far from being immovable, it has been constantly heaving and falling; and if we are not impressed by its oscillations, it is because they are not so regular or so evident to our senses as the rise and fall of the sea. The disturbances of the ocean, and the periodical advance and retreat of its tides, are known to our daily experience; we have seen it tossed into great billows by storms, or placid as a lake when undisturbed. But the crust of the earth also has had its storms, to which the tempests of the sea are as nothing,— which have thrown up mountain waves twenty thousand feet high, and fixed them where they stand, perpetual memorials of the convulsions that upheaved them. Conceive an ocean wave that should roll up for twenty thousand feet, and be petrified at its greatest height: the mountains are but the gigantic waves raised on the surface of the land by the geological tempests of past times. Besides these sudden storms of the earth's surface, there have been its gradual upheavals and depressions, going on now as steadily as ever, and which may be compared to the regular action of the tides. These, also, have had their share in determining the outlines of the continents, the height of the lands, and the depth of the seas.

Leaving aside the more general phenomena, let us look now at the formation of mountains especially. I have stated in a previous article that the relative position of the stratified and unstratified rocks gives us the key to their comparative age. To explain this I must enter into some details respecting the arrangement of stratified deposits on mountain-slopes and in mountain-chains, taking merely theoretical cases, however, to illustrate phenomena which we shall meet with repeatedly in actual facts, when studying special geological formations.

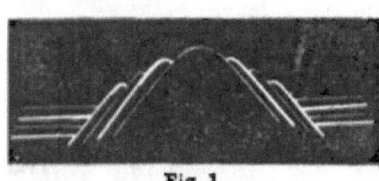
Fig. 1.

We have, for instance, in Figure 1, a central granite mountain, with a succession of stratified beds sloping against its sides, while at its base are deposited a number of horizontal beds which have evidently never been disturbed from the position in which they were originally accumulated. The reader will at once perceive the method by which the geologist decides upon the age of such a mountain. He finds the strata upon its slopes in regular superposition, the uppermost belonging, we will suppose, to the Triassic period; at its base he finds undisturbed horizontal deposits, also in regular superposition, belonging to the Jurassic and Cretaceous periods. Therefore, he argues, this mountain must have been uplifted after the Triassic and all preceding deposits were formed, since it has broken its way through them, and forced them out of their natural position; and it must have been previous to the Jurassic and Cretaceous deposits, since they have been accumulated peacefully at its base, and have undergone no such perturbations.

The task of the geologist would be an easy one, if all the problems he has to deal with were as simple as the case I have presented here; but the most cursory glance at the intricacies of mountain-structure will show us how difficult it is to trace the connection between the phenomena. We must not form an idea of ancient mountain-upheavals from existing active volcanoes, although the causes which produced them were, in a somewhat modified sense, the same. Our present volcanic mountains are only chimneys, or narrow tunnels, as it were, pierced in the thickness of the earth's surface, through which the molten lava pours out, flowing over the edges and down the sides and hardening upon the slopes, so as to form conical elevations. The mountain-ranges upheaved by ancient eruptions, on the contrary, are folds of the earth's surface, produced by the cooling of a comparatively thin crust upon a hot mass. The first effect of this cooling process would be to cause contractions; the next, to produce corresponding protrusions, — for, wherever such a shrinking and subsidence of the crust occurred, the consequent pressure upon the melted materials beneath must displace them and force them upward. While the crust continued so thin that these results could go on without very violent dislocations, — the materials within easily finding an outlet, if displaced, or merely lifting the surface without breaking through it, — the effect would be moderate elevations divided by corresponding depressions. We have seen this kind of action, during the earlier geological epochs, in the upheaval of the low hills in the United States, leading to the formation of the coal-basins.

On our return to the study of the American continent, we shall find in the Alleghany chain, occurring at a later period, between the Carboniferous and Triassic epochs, a good illustration of the same kind of phenomena, though the action of the Plutonic agents was then much more powerful, owing to the greater thickness of the crust and the consequent increase of resistance. The folds forced upward in this chain by the subsidence of the surface are higher than any preceding elevations; but they are nevertheless a succession of parallel folds divided by corresponding depressions, nor does it seem that the displacement of the materials within the crust was so violent as to fracture it extensively.

Even so late as the formation of the Jura

mountains, between the Jurassic and Cretaceous periods, the character of the upheaval is the same, though there are more cracks at right angles with the general trend of the chain, and here and there the masses below have broken through. But the chain, as a whole, consists of a succession of parallel folds, forming long domes or arches, divided by longitudinal valleys. The valleys represent the subsidences of the crust; the domes are the corresponding protrusions resulting from these subsidences. The lines of gentle undulation in this chain, so striking in contrast to the rugged and abrupt character of the Alps immediately opposite, are the result of this mode of formation.

After the crust of the earth had grown so thick, as it was, for instance, in the later Tertiary periods, when the Alps were uplifted, such an eruption could take place only through the agency of an immense force, and the extent of the fracture would be in proportion to the resistance opposed. It is hardly to be doubted, from the geological evidence already collected, that the whole mountain-range from Western Europe through the continent of Asia, including the Alps, the Caucasus, and the Himalayas, was raised at the same time. A convulsion that thus made a gigantic rent across two continents, giving egress to three such mountain-ranges, must have been accompanied by a thousand fractures and breaks in contrary directions. Such a pressure along so extensive a tract could not be equal everywhere; the various thicknesses of the crust, the greater or less flexibility of the deposits, the direction of the pressure, would give rise to an infinite variety in the results; accordingly, instead of the long, even arches, such as characterize the earlier upheavals of the Alleghanies and the Jura, there are violent dislocations of the surface, cracks, rents, and fissures in all directions, transverse to the general trend of the upheaval, as well as parallel with it.

Leaving aside for the moment the more baffling and intricate problems of the later mountain-formations, I will first endeavor to explain the simpler phenomena of the earlier upheavals.

Suppose that the melted materials within the earth are forced up against a mass of stratified deposits, the direction of the pressure being perfectly vertical, as represented in Figure 2. Such a pressure, if not too violent, would simply lift the strata out of their horizontal position into an arch or dome, (as in Figure 3,) and if contin-

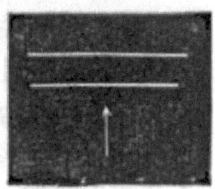

Fig. 2.

ued or repeated in immediate sequence, it would produce a number of such domes, like long billows following each other, such as we have in the

MOUNTAINS AND THEIR ORIGIN. 105

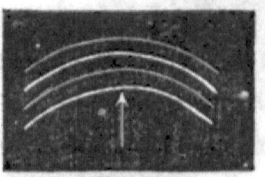
Fig. 3.

Jura. But though this is the prevailing character of the range, there are many instances even here where an unequal pressure has caused a rent at right angles with the general direction of the upheaval; and one may trace the action of this unequal pressure, from the unbroken arch, where it has simply lifted the surface into a dome, to the granite crest, where the melted rock has forced its way out and crystallized between the broken beds that rest against its slopes.

In other instances, the upper beds alone may have been cracked, while the continuity of the lower ones remains unbroken. In this case we have a longitudinal valley on the top of a mountain-range, lying between the two sides of the broken arch (as in Figure 4).

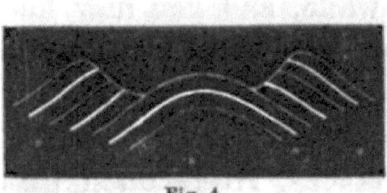
Fig. 4.

Suppose, now, that there are also transverse cracks across such a longitudinal split, we have then a longitudinal valley with transverse valleys opening into it. There are many instances of this in the Alleghanies and in the Jura. Sometimes such transverse valleys are cut straight across, so that their openings face each other; but often the cracks have taken place at different

5 *

points on the opposite sides, so that, in travelling through such a transverse valley, you turn to the right or left, as the case may be, where it enters the longitudinal valley, and follow that till you come to another transverse valley opening into it from the opposite side, through which you make your way out, thus crossing the chain in a zigzag course (as in Figure 5). Such valleys are often much narrower at some points than at others. There are even places in the Jura where a rent in the chain begins with a mere crack, — a slit but just wide enough to admit the blade of a knife; follow it for a while, and you may find it spreading gradually into a wider chasm, and finally expanding into a valley perhaps half a mile wide, or even wider.

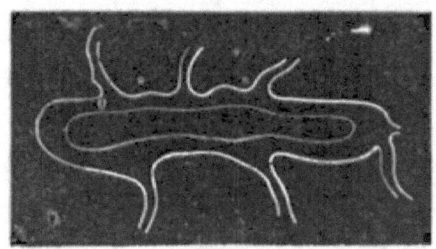

Fig. 5.

By means of such cracks, rivers often pass through lofty mountain-chains, and when we come to the investigation of the glacial phenomena connected with the course of the Rhone, we shall find that river following the longitudinal valley which separates the northern and southern parts of the chain of the Alps till it comes to Martigny, where it takes a sharp turn to the right through a transverse crack, flowing northward between walls fourteen thousand feet high, till it enters the Lake of Geneva, through which it passes, issuing at the other end, where it takes a southern direction. For a long time mountains were supposed to be the limitations of rivers, and old maps represent them always as flowing along the valleys without ever passing through the mountain-chains that divide them; but geology is fast correcting the errors of geography, and a map which represents merely the external features of a country, without reference to their structural relations, is no longer of any scientific value.

It is not, however, by rents in mountain-chains alone, or by depressions between them, that valleys are produced; they are often due to the unequal hardness of the beds raised, and to their greater or less liability to be worn away and disintegrated by the action of the rains. This inequality in the hardness of the rocks forming a mountain-range, not only adds very much to the picturesqueness of outline, but also renders the landscape more varied through the greater or less fertility of the soil. On the hard rocks, where little soil can gather, there are only pines, or a low, dwarfed growth; but on the rocks of softer materials, more easily acted upon by the rain, a richer soil gathers, and there, in the midst of mountain-scenery, may be found the most fertile growth, the richest pasturage, the brightest flowers. Where such a patch of arable soil has a southern exposure on a mountain-side, we may have a most fertile vegetation at a great height, and surrounded by the dark pine-forests. Many of the pastures on the Alps, to which from height to height the shepherds ascend with their flocks in the summer,—seeking the higher ones as the lower become dry and exhausted,—are due to such alternations in the character of the rocks.

In consequence of the influence of time, weather, atmospheric action of all kinds, the apparent relation of beds has often become so completely reversed that it is exceedingly difficult to trace their original relation. Take, for instance, the following case. An eruption has upheaved the strata over a given surface in such a manner as to lift them into a mountain, cracking open the upper beds, but leaving the lower ones unbroken. We have then a valley on a mountain-summit between two crests resembling the one already shown in Figure 4. Such a narrow passage between two crests may be changed in the course of time to a wide expansive valley by the action of the rains, frosts, and other disintegrating agents, and the relative position of the strata forming its walls may seem to be entirely changed.

Suppose, for example, that the two upper layers of the strata rent apart by the upheaval of the mountain are limestone and sandstone, while

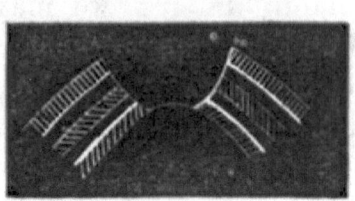
Fig. 6.

the third is clay and the fourth again limestone (as in Figure 6). Clay is soft, and yields very readily to the action of rain. In such a valley the edges of the strata forming its walls are of course exposed, and the clay formation will be the first to give way under the action of external influences. Gradually the rains wear away its substance till it is completely hollowed out. By the disintegration of the bed beneath them, the lime and sandstone layers above lose their support and crumble down, and this process goes on, the clay constantly wearing away, and the lime and sand above consequently falling in, till the upper beds have receded to a great distance, the valley has opened to a wide expanse instead of being enclosed between two walls, and the lowest

Fig. 7.

limestone bed now occupies the highest position on the mountain. Figure 7 represents one of the crests shown in Figure 6, after such a levelling process has changed its outline.

But the phenomena of eruptions in mountain-chains are far more difficult to trace than the effects thus gradually produced. Plutonic action has, indeed, played the most fantastic tricks with the crust of the earth, which seems as plastic in the grasp of the fiery power hidden within it as does clay in the hands of the sculptor.

We have seen that an equal vertical pressure from below produces a regular dome,— or that, if the dome be broken through, a granite crest is formed, with stratified materials resting against its slopes. But the pressure has often been oblique instead of vertical, and then the slope of the mountain is uneven, with a gradual ascent on one side and an abrupt wall on the other; or in some instances the pressure has been so lateral that the mountain is overturned and lies upon its side, and there are still other cases where one mountain has been tilted over and has fallen upon an adjoining one.

Sometimes, when beds have been torn asunder, one side of them has been forced up above the other; and there are even instances where one side of a mountain has been forced under the surface of the earth, while the other has remained above. Stratified beds of rock are occasionally found which have been so completely capsized, that the layers, which were of course deposited horizontally, now stand on end, side by side, in vertical rows. I remember, after a lecture on some of these extravagances in mountain-formations, a friend said to me, not inaptly, — "One can hardly help thinking of these extraordinary contortions as a succession of frantic frolics: the mountains seem like a troop of rollicking boys, hunting one another in and out and up and down in a gigantic game of hide-and-seek."

The width of the arch of a mountain depends in a great degree on the thickness and flexibility of the beds of which it is composed. There is not only a great difference in the consistency of stratified material, but every variety in the thickness of the layers, from an inch, and even less, to those measuring from ten or twenty to one hundred feet and more in depth, without marked separation of the successive beds. This is accounted for by the frequent alternations of subsidence and upheaval; the continents having tilted sometimes in one direction, sometimes in another, so that in certain localities there has been much water and large deposits, while elsewhere the water was shallow and the deposits consequently less. Thin and flexible strata have been readily lifted into a sharp, abrupt arch with narrow base, while the thick and rigid beds have been forced up more slowly in a wider arch with broader base.

Table-lands are only long unbroken folds of

the earth's surface, raised uniformly and in one direction. It is the same pressure from below, which, when acting with more intense force in one direction, makes a narrow and more abrupt fold, forming a mountain-ridge, but, when acting over a wider surface with equal force, produces an extensive uniform elevation. If the pressure be strong enough, it will cause cracks and dislocations at the edges of such a gigantic fold, and then we have table-lands between two mountain-chains, like the Gobi in Asia between the Altai Mountains and the Himalayas, or the table-land enclosed between the Rocky Mountains and the coast-range of the Pacific shore.

We do not think of table-lands as mountainous elevations, because their broad, flat surfaces remind us of the level tracts of the earth; but some of the table-lands are nevertheless higher than many mountain-chains, as, for instance, the Gobi, which is higher than the Alleghanies, or the Jura, or the Scandinavian Alps. One of Humboldt's masterly generalizations was his estimate of the average thickness of the different continents, supposing their heights to be levelled and their depressions filled up, and he found that upon such an estimate Asia would be much higher than America, notwithstanding the great mountain-chains of the latter. The extensive table-land of Asia, with the mountains adjoining it, outweighed the Alleghanies, the Rocky Mountains, the Coast-Chain, and the Andes.

When we compare the present state of our knowledge of geological phenomena with that which prevailed fifty years ago, it seems difficult to believe that so great and important a change can have been brought about in so short a time. It was on German soil and by German students that the foundation was laid for the modern science of systematic geology.

In the latter part of the eighteenth century, extensive mining operations in Saxony gave rise to an elaborate investigation of the soil for practical purposes. It was found that the rocks consisted of a succession of materials following each other in regular sequence, some of which were utterly worthless for industrial purposes, while others were exceedingly valuable. The *Muschel-Kalk* formation, so called from its innumerable remains of shells, and a number of strata underlying it, must be penetrated before the miners reached the rich veins of *Kupferschiefer* (copper slate), and below this came what was termed the *Todtliegende* (dead weight), so called because it contained no serviceable materials for the useful arts, and had to be removed before the valuable beds of coal lying beneath it, and making the base of the series, could be reached. But while the workmen wrought at these successive layers of rock to see what they would yield for practical purposes, a man was watching their operations who considered the crust of the earth from quite another point of view.

Abraham Gottlob Werner was born more than a century ago in Upper Lusatia. His very infancy seemed to shadow forth his future studies, for his playthings were the minerals he found in his father's forge. At a suitable age he was placed at the mining school of Freiberg in Saxony, and having, when only twenty-four years of age, attracted attention in the scientific world by the publication of an "Essay on the Characters of Minerals," he was soon after appointed to the professorship of mineralogy in Freiberg. His lot in life could not have fallen in a spot more advantageous for his special studies, and the enthusiasm with which he taught communicated itself to his pupils, many of whom became his devoted disciples, disseminating his views in their turn with a zeal which rivalled the master's ardor.

Werner took advantage of the mining operations going on in his neighborhood, the blasting, sinking of shafts, etc., to examine critically the composition of the rocks thus laid open, and the result of his analysis was the establishment of the Neptunic school of geology alluded to in a previous article, and so influential in science at the close of the eighteenth and the opening of the nineteenth century. From the general character of these rocks, as well as the number of marine shells contained in them, he convinced himself that the whole series, including the Coal, the *Todtliegende*, the *Kupferschiefer*, the *Zechstein*, the Red Sandstone, and the *Muschel-Kalk*, had been deposited under the agency of water, and were the work of the ocean.

Thus far he was right, with the exception that he did not include the accumulation of materials by the local action of fresh water afterwards traced by Cuvier and Brogniart in the Tertiary deposits about Paris. But from these data he went a step too far, and assumed that all rocks, except the modern lavas, must have been accumulated by the sea, — believing even the granites, porphyries, and basalts to have been deposited in the ocean and crystallized from the substances it contained in solution.

But, in the mean time, James Hutton, a Scotch geologist, was looking at phenomena of a like character from a very different point of view. In the neighborhood of Edinburgh, where he lived, was an extensive region of trap-rock, — that is, of igneous rock, which had forced itself through the stratified deposits, sometimes spreading in a continuous sheet over large tracts, or splitting them open and filling all the interstices and cracks so formed. Thus he saw igneous rocks not only covering or underlying stratified deposits, but penetrating deep into their structure, forming dikes at right angles with them, and presenting, in short, all the phenomena belonging to volcanic rocks in contact with stratified materials. He again pushed his theory too far, and, inferring from the phenomena immediately about him that heat had been the chief agent in the formation of the earth's crust, he was inclined to believe that the stratified materials also were in part at least due to this cause. I have alluded in a former number to the hot disputes and long-contested battles of geologists upon this point. It was a pupil of Werner's who at last set at rest this much vexed question.

At the age of sixteen, in the year 1790, Leopold von Buch was placed under Werner's care at the mining school of Freiberg. Werner found him a pupil after his own heart. Warmly adopting his teacher's theory, he pursued his geological studies with the greatest ardor, and continued for some time under the immediate influence and guidance of the Freiberg professor. His university studies over, however, he began to pursue his investigations independently, and his geological excursions led him into Italy, where his confidence in the truth of Werner's theory began to be shaken. A subsequent visit to the region of extinct volcanoes in Auvergne, in the South of France, convinced him that the aqueous theory was at least partially wrong, and that fire had been an active agent in the rock-formations of past times. This result did not change the convictions of his master, Werner, who was too old or too prejudiced to accept the later views, which were nevertheless the result of the stimulus he himself had given to geological investigations.

But Von Buch was indefatigable. For years he lived the life of an itinerant geologist. With a shirt and a pair of stockings in his pocket, and a geological hammer in his hand, he travelled all over Europe on foot. The results of his foot-journey to Scandinavia were among his most important contributions to geology. He went also to the Canary Islands; and it is in his extensive work on the geological formations of these islands that he showed conclusively not only the Plutonic character of all unstratified rocks, but also that to their action upon the stratified deposits the inequalities of the earth's surface are chiefly due. He first demonstrated that the melted masses within the earth had upheaved the materials deposited in layers upon its surface, and had thus formed the mountains.

No geologist has ever collected a larger amount of facts than Von Buch, and to him we owe a great reform not only in geological principles, but in methods of study also. An amusing anecdote is told of him, as illustrating his untiring devotion to his scientific pursuits. In studying the rocks, he had become engaged also in the investigation of the fossils contained in them. He was at one time especially interested in the *Terebratulæ*, certain fossil shells found in great abundance in all stratified rocks, and one evening in Berlin, where he was engaged in the study of these remains, he came across a notice in a Swedish work of a particular species of that family which he could not readily identify without seeing the original specimens. The next morning Von Buch was missing, and as he had invited guests to dine with him, some anxiety was felt on account of his non-appearance. On inquiry, it was found that he was already far on his way to Sweden: he had started by daylight on a pilgrimage after the new, or rather the old, *Terebratula*. I tell the story as I heard it from one of the disappointed guests.

All great natural phenomena impressed him deeply. On one occasion it was my good fortune to make one of a party from the "Helvetic Association for the Advancement of Science" on an excursion to the eastern extremity of the Lake of Geneva. I well remember the expressive gesture of Von Buch, as he faced the deep gorge through which the Rhone issues from the interior of the

Alps. While others were chatting and laughing about him, he stood for a moment absorbed in silent contemplation of the grandeur of the scene, then lifted his hat and bowed reverently before the mountains.

Next to Von Buch, no man has done more for modern geology than Elie de Beaumont, the great French geologist. Perhaps the most important of his generalizations is that by which he has given us the clew to the limitation of the different epochs in past times by connecting them with the great revolutions in the world's history. He has shown us that the great changes in the aspect of the globe, as well as in its successive sets of animals, coincide with the mountain-upheavals.

I might add a long list of names, American as well as European, which will be forever honored in the history of science for their contributions to geology in the last half-century. But I have intended only to close this chapter on mountains with a few words respecting the men who first investigated their intimate structural organization, and established methods of study in reference to them now generally adopted throughout the scientific world. In my next article I shall proceed to give some account of special geological formations in Europe, and the gradual growth of that continent.

V.

THE GROWTH OF CONTINENTS.

BEFORE entering upon a sketch of the growth of the European Continent from the earliest times until it reached its present dimensions and outlines, I will say something of the growth of continents in general, connecting these remarks with a few words of explanation respecting some geological terms, which, although in constant use, are nevertheless not clearly defined. I will explain, at the outset, the meaning I attach to them and the sense in which I use them, that there may be no misunderstanding between me and my readers on this point. The words Age, Epoch, Period, Formation, may be found on almost every page of any modern work on geology; but if we sift the matter carefully, we shall find that there is a great uncertainty as to the significance of these terms, and that scarcely any two geologists use them in the same sense. Indeed, I shall not be held blameless in this respect myself; for, on looking over preceding articles, I find that I have, from old habit, used somewhat indiscriminately names which should have a perfectly definite and invariable meaning. As long as zoölogical nomenclature was uncontrolled by any principle, the same vagueness and indecision prevailed here also. The words Genus, Order, Class, as well as those applied to the most comprehensive division of all in the animal kingdom, the primary branches or types, were used indiscriminately, and often allowed to include under one name animals differing essentially in their structural character. It is only since it has been found that all these groups are susceptible of limitation, according to distinct categories of structure, that our nomenclature has assumed a more precise and definite significance. Even now there is still some inconsistency among zoölogists as to the use of special terms, arising from their individual differences in appreciating structural features; but I believe it to be, nevertheless, true, that genera, orders, classes, etc., are not merely larger or smaller groups of the same kind, but are really based upon distinct categories of structure. As soon as such a principle is admitted in geology, and investigators recognize certain physical and organic conditions, more or less general in their action, as characteristic of all those chapters in geological history designated as Ages, Epochs, Periods, Formations, etc., all vagueness will vanish from the scientific nomenclature of this department also, and there will be no hesitation as to the use of words for which we shall then have a positive, definite meaning.

Although the fivefold division of Werner, by which he separated the rocks into Primitive, Transition, Secondary, Alluvial, and Volcanic, proved to be based on a partial misapprehension of the nature of the earth-crust, yet it led to their subsequent division into the three great groups now known as the Primary, or Palæozoic, as they are sometimes called, because here are found the first organic remains, the Secondary, and the Tertiary. I have said in a previous article that the general unity of character prevailing throughout these three divisions, so that, taken from the broadest point of view, each one seems a unit in time, justifies the application to them of that term, *Age*, by which we distinguish in human history those periods marked throughout by one prevailing tendency;— as we say the age of Egyptian or Greek or Roman civilization,— the age of stone or iron or bronze. I believe that this division of geological history into these great sections or chapters is founded upon a recognition of the general features by which they are characterized.

Passing over the time when the first stratified deposits were accumulated under a universal ocean in which neither animals nor plants existed, there was an age in the physical history of the world when the lands consisted of low islands, — when neither great depths nor lofty heights diversified the surface of the earth, — when both the animal and vegetable creation, however numerous, was inferior to the later ones, and comparatively uniform in character, — when marine Cryptogams were the highest plants, and Fishes were the highest animals. And this broad statement holds good for the whole of that time, even though it was not without its minor changes, its new forms of animal and vegetable life, its variations of level, its upheavals and subsidences; for, nevertheless, through its whole duration, it was the age of low detached lands, — it was the age of Cryptogams, — it was the age of Fishes. From its beginning to its close, no higher type in the animal kingdom, no loftier group in the vegetable world, made its appearance.

There was an age in the physical history of the world when the patches of land already raised above the water became so united as to form large islands; and though the aspect of the earth retained its insular character, yet the size of the islands, their tendency to coalesce by the addition of constantly increasing deposits, and thus to spread into wider expanses of dry land, marked the advance toward the formation of continents. This extension of the dry land was brought about not only by the gradual accumulation of materials, but also by the upheaval of large tracts of stratified deposits; for, though the loftiest mountain-chains did not yet exist, ranges like those of the Alleghanies and the Jura belong to this division of the world's history. During this time, the general character of the animal and vegetable kingdoms was higher than during the previous age. Reptiles, many and various, gigantic in size, curious in form, some of them recalling the structure of fishes, others anticipating birdlike features, gave a new character to the animal world, while in the vegetable world the reign of the aquatic Cryptogams was over, and terrestrial Cryptogams, and, later, Gymnosperms and Monocotyledonous trees, clothed the earth with foliage. Such was the character of this second age, from its opening to its close; and though there are indications that, before it was wholly past, some low, inferior Mammalian types of the Marsupial kind were introduced,* and also a few Dicotyledonous plants, yet they were not numerous or

* I say nothing of the traces of Birds in the Secondary deposits, because the so-called bird-tracks seem to me of very doubtful character; and it is also my opinion that the remains of a feathered animal recently found in the Solenhofen lithographic limestone, and believed to be a bird by some naturalists, do not belong to a genuine bird, but to one of those synthetic types before alluded to, in which reptilian structure is combined with certain birdlike features.

striking enough to change the general aspect of the organic world. This age was throughout, in its physical formation, the age of large continental islands; while in its organic character it was the age of Reptiles as the highest animal type, and of Gymnosperms and Monocotyledonous plants as the highest vegetable groups.

There was an age in the physical history of the world when great ranges of mountains bound together in everlasting chains the islands which had already grown to continental dimensions,— when wide tracts of land, hitherto insular in character, became soldered into one by the upheaval of Plutonic masses which stretched across them all and riveted them forever with bolts of granite, of porphyry, and of basalt. Thus did the Rocky Mountains and the Andes bind together North and South America; the Pyrenees united Spain to France; the Alps, the Caucasus, and the Himalayas bound Europe to Asia. The class of Mammalia was now at the head of the animal kingdom; huge quadrupeds possessed the earth, and dwelt in forests characterized by plants of a higher order than any preceding ones,— the Beeches, Birches, Maples, Oaks, and Poplars of the Tertiaries. But though the continents had assumed their permanent outlines, extensive tracts of land still remained covered with ocean. Inland seas, sheets of water like the Mediterranean, so unique in our world, were then numerous. Physically speaking, this was the age of continents broken by large inland seas; while in the organic world it was the age of Mammalia among animals, and of extensive Dicotyledonous forests among plants. In a certain sense it was the age of completion, — the one which ushered in the crowning work of creation.

There was an age in the physical history of the world (it is in its infancy still) when Man, with the animals and plants that were to accompany him, was introduced upon the globe, which had acquired all its modern characters. At last the continents were redeemed from the water, and all the earth was given to this new being for his home. Among all the types born into the animal kingdom before, there had never been one to which positive limits had not been set by a law of geographical distribution absolutely impassible to all. For Man alone those boundaries were removed. He, with the domestic animals and plants which were to be the companions of all his pilgrimages, could wander over the whole earth and choose his home. Placed at the head of creation, gifted with intellect to make both animals and plants subservient to his destinies, his introduction upon the earth marks the last great division in the history of our planet. To designate these great divisions in time, I would urge, for the reasons above stated, that the term which is indeed often, though not invariably, applied to them, be exclusively adopted,—that of the Ages of Nature.

But these Ages are themselves susceptible of subdivisions, which should also be accurately defined. What is the nature of these subdivisions? They are all connected with sudden physical changes in the earth's surface, more or less limited in their action, these changes being themselves related to important alterations in the organic world. Although I have stated that one general character prevailed during each of the Ages, yet there was nevertheless a constant progressive action running through them all, and at various intervals both the organic and the physical world received a sudden impulse in consequence of marked and violent changes in the earth-crust, bringing up new elevations, while at the same time the existing animal creation was brought to a close, and a new set of beings was introduced. These changes are not yet accurately defined in America, because the age of her mountains is not known with sufficient exactness; but their limits have been very extensively traced in Europe, and this coincidence of the various upheavals with the introduction of a new population differing entirely from the preceding one, has been demonstrated so clearly, that it may be considered as an ascertained law. What name, then, is most appropriate for the divisions thus marked by sudden and violent changes? It seems to me, from their generally accepted meaning, that the word Epoch or Era, both of which have been widely, though indiscriminately, used in geology, is especially applicable here. In their common use, they imply a condition of things determined by some decisive event. In speaking of human affairs, we say, " It was an epoch or an era in history,"— or in a more limited sense, " It was an epoch in the life of such or such a man." It at once conveys the idea of an important change connected with or brought about by some striking occurrence. Such were those divisions in the history of the earth when a violent convulsion in the surface of the globe and a change in its inhabitants ushered in a new aspect of things.

I have said that we owe to Elie de Beaumont the discovery of this connection between the successive upheavals and the different sets of animals and plants which have followed each other on the globe. We have seen, in the preceding article upon the formation of mountains, that the dislocations thus produced show the interruptions between successive deposits: as, for instance, where certain strata are raised upon the sides of a mountain, while other strata rest *unconformably*, as it is called, above them at its base,— this term, unconformable, signifying merely that the two sets of strata are placed at an entirely different angle, and must therefore belong to two distinct sets of deposits. But there are two series of geological facts connected with this result which are often confounded, though they arise from very different causes. One is that described above, in which, a certain series of beds having been raised out of their natural horizontal position, another series has been deposited upon them, thus resting unconformably above. The other is where, one set of beds having been deposited over any given region, at a later time, in consequence of a recession of the sea-shore, for instance, or of some other gradual disturbance of the surface, the next set of beds accumulated above them cover a somewhat different area, and are therefore not conformable with the first, though parallel with them. This difference, however slight, is sufficient to show that some shifting of the ground on which they were accumulated must have taken place between the two series of deposits.

This distinction must not be confounded with that made by Elie de Beaumont: we owe it to D'Orbigny, who first pointed out the importance of distinguishing the dislocations produced by gradual movements of the earth from those caused by mountain-upheavals. The former are much more numerous than the latter, and in every epoch geologists have distinguished a number of such changes in the surface of the earth, accompanied by the introduction of a new set of animals, though the changes in the organic world are not so striking as those which coincide with the mountain-upheavals. Still, to the eye of the geologist they are quite as distinct, though less evident to the ordinary observer. To these divisions it seems to me that the name of Period is rightly applied, because they seem to have been brought about by the steady action of time, and by gradual changes, rather than by any sudden or violent convulsion.

It was my good fortune to be in some degree connected with the investigations respecting the limitation of Periods, for which the geology of Switzerland afforded peculiar facilities. My early home was near the foot of the Jura, where I constantly faced its rounded domes and the slope by which they gently descend to the plain of Switzerland. I have heard it said that there is something monotonous in the continuous undulations of this range, so different from the opposite one of the Alps. But I think it is only by contrast that it seems wanting in vigor and picturesqueness; and those who live in its neighborhood become very much attached to the more peaceful character of its scenery. Perhaps my readers will pardon the digression, if I interrupt our geological discussion for a moment, to offer them a word of advice, though it be uncalled for. I have often been asked by friends who were intending to go to Europe, what is the most favorable time in the day and the best road to enter Switzerland in order to have at once the finest impression of the mountains. My answer is always, — Enter it in the afternoon over the Jura. If you are fortunate, and have one of the bright, soft afternoons that sometimes show the Alps in their full beauty, as you descend the slope of the Jura, from which you command the whole panorama of the opposite range, you may see, as the day dies, the last shadow pass with strange rapidity from peak to peak of the Alpine summits. The passage is so rapid, so sudden, as the shadow vanishes from one height and appears on the next, that it seems like the step of some living spirit of the mountains. Then, as the sun sinks, it sheds a brilliant glow across them, and upon that follows, — strangest effect of all, — a sudden pallor, an ashy paleness on the mountains, that has a ghastly, chilly look. But this is not their last aspect: after the sun has vanished out of sight, in place of the glory of his departure, and of the corpse-like pallor which succeeded it, there spreads over the mountains a faint blush that dies gradually into the night. These changes,— the glory, the death, the soft succeeding life,— really seem like something that has a spiritual existence. While, however, I counsel my friends to see the Alps for the first time in the afternoon, if possible, I do not promise them that the hour will bring with it such a scene as I have tried to describe. Perfect sunsets are rare in any land; but, nevertheless, I would advise travellers to choose the latter half of the day and a road over the Jura for their entrance into Switzerland.*

It was from the Jura itself that one of the great epochs in the history of the globe received its name. It was in a deep gorge of the Jura, that, more than half a century ago, Leopold von Buch first perceived the mode of formation of mountains; and it was at the foot of the Jura,

---

* The two most imposing views of the Alps from the Jura are those of Latourne, on the road from Pontarlier to Neufchatel, and of St. Cergues, on the road from Lons le Saulnier to Nyon; the next best is to be had above Boujean, on the road from Basle to Bienne. Very extensive views may be obtained from any of the summits in the southern range of the Jura; among which the Weissenstein above Soleure, the Chasseral above Bienne, the Chaumont above Neufchatel, the Chasseron above Grançon, the Suchet above Orbe, the Mont Tendre or the Noirmont above Morges, and the Dôle above Nyon, are the most frequented. Of all these points Chaumont is unquestionably to be preferred, as it commands at the same time an equally extensive view of the Bernese Alps and the Mont Blanc range.

in the neighborhood of Neufchatel, that the investigations were made which first led to the recognition of the changes connected with the Periods. As I shall have occasion hereafter to enter into this subject more at length, I will only allude briefly here to the circumstances. In so doing I am anticipating the true geological order, because I must treat of the Jurassic and Cretaceous deposits, which are still far in advance of us; but as it was by the study of these deposits that the circumscription of the Periods, as I have defined them above, was first ascertained, I must allude to them in this connection.

Facing the range of the Jura from the Lake of Neufchatel, there seems to be but one uninterrupted slope by which it descends to the shore of the lake. It will, however, be noticed by the most careless observer that this slope is divided by the difference in vegetation into two strongly marked bands of color: the lower and more gradual descent being of a lighter green, while the upper portion is covered by the deeper hue of the forest-trees, the Beeches, Birches, Maples, etc., above which come the Pines. When the vegetation is fully expanded, this marked division along the whole side of the range into two broad bands of green, the lighter below and the darker above, becomes very striking. The lighter band represents the cultivated portion of the slope, the vineyards, the farms, the orchards, covering the gentler, more gradual part of the descent; and the whole of this cultivated tract, stretching a hundred miles east and west, belongs to the Cretaceous epoch. The upper slope of the range, where the forest-growth comes in, is Jurassic. Facing the range, you do not, as I have said, perceive any difference in the angle of inclination; but the border-line between the two bands of green does in fact mark the point at which the Cretaceous beds abut with a gentler slope against the Jurassic strata, which continue their sharper descent, and are lost to view beneath them.

This is one of the instances in which the contact of two epochs is most directly traced. There is no question, from the relation of the deposits, that the Jura in its upheaval carried with it the strata previously accumulated. At its base there was then no lake, but an extensive stretch of ocean; for the whole plain of Switzerland was under water, and many thousand years elapsed before the Alps arose to set a new boundary to the sea and enclose that inland sheet of water, gradually to be filled up by more modern accumulations, and transformed into the fertile plain which now lies between the Jura and the Alps. If the reader will for a moment transport himself in imagination to the time when the southern side of the Jurassic range sloped directly down to the ocean, he will easily understand how this second series of deposits was collected at its base, as materials are collected now along any sea-shore. They must of course have been accumulated horizontally, since no loose materials could keep their place even at so moderate an angle as that of the present lower slope of the range; but we shall see hereafter that there were many subsequent perturbations of this region, and that these Cretaceous deposits, after they had become consolidated, were raised by later upheavals from their original position to that which they now occupy on the lower slope of the Jura, resting immediately, but in geological language *unconformably*, against it. The two adjoining wood-cuts are merely theoretical, showing by lines the past and the present relation of these deposits; but they may assist the reader to understand my meaning.

Figure 1 represents the Jura before the Alps were raised, with the Cretaceous deposits accu-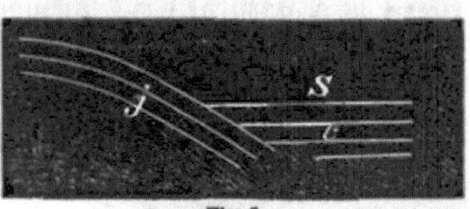

Fig. 1.

mulating beneath the sea at its base. The line marked S indicates the ocean-level; the letter c, the Cretaceous deposits; the letter j, the Jurassic strata, lifted on the side of the mountain.

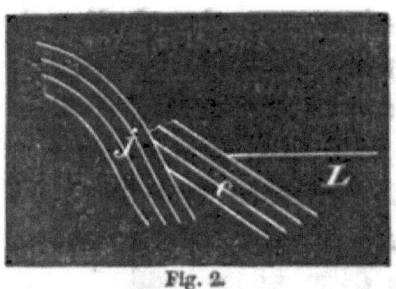

Fig. 2.

Figure 2 represents the Jura at the present time, when the latter upheavals have lifted the Jurassic strata to a sharper inclination with the Cretaceous deposits, now raised and forming the lower slope of the mountain, at the base of which is the Lake of Neufchatel, marked *L* in the diagram.

Although this change of inclination is hardly perceptible, as one looks up against the face of the Jura range, there is a transverse cut across it which seems intended to give us a diagram of its internal structure. Behind the city of Neufchatel rises the mountain of Chaumont, so called from its bald head, for neither tree nor shrub grows on its summit. Straight through this mountain, from its northern to its southern side, there is a natural road, formed by a split in the mountain from top to bottom. In this transverse cut, which forms one of the most romantic and picturesque gorges leading into the heart of the Jura range, you get a profile view of the change in the inclination of the strata, and can easily distinguish the point of juncture between the two sets of deposits. But even after this dislocation of strata had been perceived, it was not known that it indicated the commencement of a new epoch, and it is here that my own share in the work, such as it is, belongs.

Accustomed as a boy to ramble about in the beautiful gorges and valleys of the Jura, and in riper years, as my interest in science increased, to study its formation with closer attention, this difference in the inclination of the slope had not escaped my observation. I was, however, still more attracted by the fossils it contained than by its geological character: and, indeed, there is no better locality for the study of extinct forms of life than the Jura. In all its breaks and ravines, wherever the inner surface of the rock is exposed, it is full of organic remains; and to take a handful of soil from the roadside is often to gather a handful of shells. It is actually built of the remains of animals, and there are no coral reefs in existing seas presenting a better opportunity for study to the naturalist than the coral reefs of the Jura. Being already tolerably familiar with the fossils of the Jura, it occurred to me to compare those of the upper and lower slope; and to my surprise I found that they were everywhere different, and that those of the lower slope were invariably Cretaceous in character, while those of the upper slope were Jurassic. In the course of this investigation I discovered three periods in the Cretaceous and four in the Jurassic epoch, all characterized by different fossils. This led to a more thorough investigation of the different sets of strata, resulting in the establishment by D'Orbigny of a still greater number of periods, marked by the successive deposits of the Jurassic and Cretaceous seas, all of which contained different organic remains. The attention of geologists being once turned in this direction, the other epochs were studied with the same view, and all were found to be susceptible of division into a greater or less number of such periods.

I have dwelt at greater length on the Jurassic and Cretaceous divisions, because I believe that we have in the relation of these two epochs, as well as in that of the Cretaceous epoch with the Tertiary immediately following it, facts which are very important in their bearing on certain questions, now loudly discussed, not only by scientific men, but by all who are interested in the mode of origin of animals. Certainly, in the inland seas of the Cretaceous and subsequent Tertiary times, where we can trace in the same sheet of water not only the different series of deposits belonging to two successive epochs in immediate juxtaposition, but those belonging to all the periods included within these epochs, with the organic remains contained in each,— there, if anywhere, we should be able to trace the transition-types by which one set of animals is said to have been developed out of the preceding. We hear a great deal of the interruption in geological deposits, of long intervals, the record of which has vanished, and which may contain those intermediate links for which we vainly seek. But here there is no such gap in the evidence. In the very same sheets of water, covering limited areas, we have the successive series of deposits containing the remains of animals which continue perfectly unchanged during long intervals. Immediately upon these, and accompanied by a more or less violent shifting of the surface,* traceable by the consequent discordance of the strata, is introduced an entirely new set of animals, differing as much from those immediately preceding them as do those of the present period from the older animals, (our predecessors, but *not* our ancestors,) traced by Cuvier in the Tertiary deposits underlying those of our own geological age. I subjoin here a tabular view giving the Epochs in their relation to the Ages, and indicating, at least approximately, the number of Periods contained in each Epoch.

Age of Man	Present Epoch.	
Tertiary Age: Age of Mammalia	{ Pliocene Miocene Eocene }	with at least twelve Periods.
Secondary Age: Age of Reptiles	{ Cretaceous Jurassic Triassic }	with about twenty Periods.
	{ Permian Carboniferous }	with eight or nine Periods.
Palæozoic or Primary Age: Age of Fishes	{ Devonian Silurian }	with ten or twelve Periods.

* I use surface often in its geological significance, meaning earth-crust, and applied to sea-bottom as well as to dry land.

It will be noticed by those who have any knowledge of geological divisions, that in this diagram I consider the Carboniferous epoch as forming a part of the Secondary age. Some geologists have been inclined, from the marked and peculiar character of its vegetation, to set it apart as forming in itself a distinct geological age, while others have united it with the Palæozoic age. For many years I myself adopted the latter of these two views, and associated the Carboniferous epoch with the Palæozoic age. But it is the misfortune of progress that one is forced not only to unlearn a great deal, but, if one has been in the habit of communicating his ideas to others, to destroy much of his own work. I now find myself in this predicament; and after teaching my students for years that the Carboniferous epoch belongs to the Palæozoic or Primary age, I am convinced, — and this conviction grows upon me constantly as I free myself from old prepossessions and bias on the subject, — that with the Carboniferous epoch we have the opening of the Secondary age in the history of the world. A more intimate acquaintance with organic remains has shown me that there is a closer relation between the character of the animal and vegetable world of the Carboniferous epoch, as compared with that of the Permian and Triassic epochs, than between that of the Carboniferous epoch and any preceding one.

Neither do I see any reason for separating it from the others as a distinct age. The plants as well as the animals of the two subsequent epochs seem to me to show, on the contrary, the same pervading character, indicating that the Carboniferous epoch makes an integral part of that great division which I have characterized as the Secondary age.

Within the Periods there is a still more limited kind of geological division, founded upon the special character of local deposits. These I would call geological Formations, indicating concrete local deposits, having no cosmic character, but circumscribed within comparatively narrow areas, as distinguished from the other terms, Ages, Epochs, Periods, which have a more universal meaning, and are, as it were, cosmopolitan in their application. Let me illustrate my meaning by some formations of the present time. The accumulations along the coast of Florida are composed chiefly of coral sand, mixed of course with the remains of the animals belonging to that locality; those along the coast of the Southern States consist principally of loam, which the rivers bring down from their swamps and low, muddy grounds; those upon the shores of the Middle States are made up of clay from the disintegration of the eastern slopes of the Alleghanies; while those farther north, along our own coast, are mostly formed of sand from the New-England granites. Such deposits are the local work of one period, containing the organic remains belonging to the time and place. From the geological point of view, I would call them Formations; from the naturalist's point of view I would call them Zoölogical Provinces.

Of course, in urging the application of these names, I do not intend to assume any dictatorship in the matter of geological nomenclature. But I do feel very strongly the confusion arising from an indiscriminate use of terms, and that, whatever names be selected as most appropriate or descriptive for these divisions, geologists should agree to use them in the same sense.

There is one other geological term, bequeathed to us by a great authority, and which cannot be changed for the better: I mean that of Geological Horizon, applied by Humboldt to the whole extent of any one geological division,—as, for instance, the Silurian horizon, including the whole extent of the Silurian epoch. It indicates one level in time, as the horizon which limits our view indicates the farthest extension of the plain on which we stand in space.

We left America at the close of the Carboniferous epoch, when the central part of the United States was already raised above the water. Let us now give a glance at Europe in those early days, and see how far her physical history has advanced. What European countries loom up for us out of the Azoic sea, corresponding in time and character to the low range of hills which first defined the northern boundary of the United States? what did the Silurian and Devonian epochs add to these earliest tracts of dry land in the Old World? and where do we find the coal basins which show us the sites of her Carboniferous forests? Since the relation between the epochs of comparative tranquillity and the successive upheavals has been so carefully traced in Europe, I will endeavor, while giving a sketch of that early European world, to point out, at the same time, the connection of the different systems of upheaval with the successive stratified deposits, without, however, entering into such details as must necessarily become technical and tedious.

In the European ocean of the Azoic epoch we find five islands of considerable size. The largest of these is at the North. Scandinavia had even then almost her present outlines; for Norway, Sweden, Finland, and Lapland, all of which are chiefly granitic in character, were among the first lands to be raised. Between Sweden and Norway there is, however, still a large tract of land under water, forming an extensive lake or a large inland sea in the heart of the country. If the reader will take the trouble to look on any geological map of Europe, he will see an extensive patch of Silurian rock in the centre of Sweden and Norway. This represents that sheet of water gradually to be filled by the accumulation of Silurian deposits and afterwards raised by a later disturbance. There is another mass of land far to the southeast of this Scandinavian island, which we may designate as the Bohemian island, for it lies in the region now called Bohemia, though it includes, also, a part of Saxony and Moravia. The northwest corner of France, that promontory which we now call Bretagne, with a part of Normandy adjoining it, formed another island; while to the southeast of it lay the central plateau of France. Great Britain was not forgotten in this early world; for a part of the Scotch hills, some of the Welsh mountains, and a small elevation here and there in Ireland, already formed a little archipelago in that region. By a most careful analysis of the structure of the rocks in these ancient patches of land, tracing all the dislocations of strata, all the indications of any disturbance of the earth-crust whatsoever, Elie de Beaumont has detected and classified four systems of upheavals, previous to the Silurian epoch, to which he refers these islands in the Azoic sea. He has named them the systems of

La Vendée, of Finistère, of Longmynd, and of Morbihan. These names have, for the present, only a local significance,—being derived, like so many of the geological names, from the places where the investigations of the phenomena were first undertaken; but in course of time they will, no doubt, apply to all the contemporaneous upheavals, wherever they may be traced, just as we now have Silurian, Devonian, Permian, and Jurassic deposits in America as well as in Europe.

The Silurian and Devonian epochs seem to have been instrumental rather in enlarging the tracts of land already raised than in adding new ones; yet to these two epochs is traced the upheaval of a large and important island to the northeast of France. We may call it the Belgian island, since it covered the ground of modern Belgium; but it also extended considerably beyond these limits, and included much of the Northern Rhine region. A portion only of this tract, to which belongs the central mass of the Vosges and the Black Forest, was lifted during the Silurian epoch,—which also enlarged considerably Wales and Scotland, the Bohemian island, the island of Bretagne, and Scandinavia. During this epoch the sheet of water between Norway and Sweden became dry land, a considerable tract was added to their northern extremity on the Arctic shore; while a broad band of

Silurian deposits, lying now between Finland and Russia, enlarged that region.

The Silurian epoch has been referred by Elie de Beaumont to the system of upheaval called by him the system of Westmoreland and Hundsrück,—again merely in reference to the spots at which these upheavals were first studied, the centres, as it were, from which the investigations spread. But in their geological significance they indicate all the oscillations and disturbances of the soil throughout the region over which the Silurian deposits have been traced in Europe. The Devonian epoch added greatly to the outlines of the Belgian island. To it belongs the region of the Ardennes, lying between France and Belgium, the Eifelgebirge, and a new disturbance of the Vosges, by which that region was also extended. The island of Bretagne was greatly increased by the Devonian deposits, and Bohemia gained in dimensions, while the central plateau of France remained much the same as before. The changes of the Devonian epoch are traced by Elie de Beaumont to a system of upheavals called the Ballons of the Vosges and of Normandy,—so called from the rounded, balloon-like domes characteristic of the mountains of that time. To the Carboniferous epoch belong the mountain-systems of Forey, (to the west of Lyons,) of the North of England, and of the

Netherlands. These three systems of upheaval have also been traced by Elie de Beaumont; and in the depressions formed between their elevations we find the coal-basins of Central France, of England, and of Germany. During all these epochs, in Europe as in America, every such dislocation of the surface was attended by a change in the animal creation.

If we take now a general view of the aspect of Europe at the close of the Carboniferous epoch, we shall see that the large island of Scandinavia is completed, while the islands of Bohemia and Belgium have approached each other by their gradual increase till they are divided only by a comparatively narrow channel. The island of Belgium, that of Bretagne, and that of the central plateau of France, form together a triangle, of which the plateau is the lowest point, while Belgium and Bretagne form the other two corners. Between the plateau and Belgium flows a channel, which we may call the Burgundian channel, since it covers old Burgundy; between the plateau and Bretagne is another channel, which from its position we may call the Bordeaux channel. The space inclosed between these three masses of land is filled by open sea. To trace the gradual closing of these channels and the filling up of the ocean by constantly increasing accumulations, as well as by upheavals, will be the object of the next article.

VI.

THE GEOLOGICAL MIDDLE AGE.

I SHALL pass lightly over the Permian and Triassic epochs, as being more nearly related in their organic forms to the Carboniferous epoch, with which we are already somewhat familiar, while in those next in succession, the Jurassic and Cretaceous epochs, the later conditions of animal life begin to be already foreshadowed. But though less significant for us in the present stage of our discussion, it must not be supposed that the Permian and Triassic epochs were unimportant in the physical and organic history of Europe. A glance at any geological map of Europe will show the reader how the Belgian island stretched gradually in a southwesterly direction during the Permian epoch, approaching the coast of France by slowly increasing accumulations, and thus filling the Burgundian channel; a wide border of Permian deposits around the coal-field of Great Britain marks the increase of this region also during the same time, and a very extensive tract of a like character is to be seen in

Russia. The latter is, however, still under doubt and discussion among geologists, and more recent investigations tend to show that this Russian region, supposed at first to be exclusively Permian, is in part at least, Triassic.

With the coming in of the Triassic epoch began the great deposits of Red Sandstone, Muschel-Kalk, and Keuper, in Central Europe. They united the Belgian island to the region of the Vosges and the Black Forest, while they also filled to a great extent the channel between Belgium and the Bohemian island. Thus the land slowly gained upon the Triassic ocean, shutting it within ever-narrowing limits, and preparing the large inland seas so characteristic of the later Secondary times.

The character of the organic world still retained a general resemblance to that of the Carboniferous epoch. Among Radiates, the Corals were more nearly allied to those of the earlier ages than to those of modern times, and Crinoids abounded still, though some of the higher Echinoderm types were already introduced. Among Mollusks, the lower Bivalves, that is, the Brachiopods and Bryozoa, still prevailed, while Ammonites continued to be very numerous, differing from the earlier ones chiefly in the ever-increasing complications of their inner partitions, which become so deeply involuted and cut upon their margins, before the type disappears, as to make an intricate tracery of very various patterns on the surface of these shells. The most conspicuous type of Articulates continues as before to be that of Crustacea; but Trilobites have finished their career, and the Lobster-like Crustacea make their appearance for the first time. It does not seem that the class of Insects has greatly increased since the Carboniferous epoch; and Worms are still as difficult to trace as ever, being chiefly known by the cases in which they sheltered themselves. Among Vertebrates, the Fishes still resemble those of the Carboniferous epoch, belonging principally to the Selachians and Ganoids. They have, however, approached somewhat toward a modern pattern, the lobes of the tail being more evenly cut, and their general outline more like that of common fishes. The gigantic marsh Reptiles have become far more numerous and various. They continue through several epochs, but may be said to reach their culminating point in the Jurassic and Cretaceous deposits.

I cannot pass over the Triassic epoch without some allusion to the so-called bird-tracks, so generally believed to mark the introduction of Birds at this time. It is true that in the deposits of the Trias there have been found many traces of footsteps, indicating a vast number of animals which, except for these footprints, remain unknown to us. In the sandstone of the Connecticut Valley they are found in extraordinary numbers, as if these animals, whatever they were, had been in the habit of frequenting that shore. They appear to have been very diversified; for some of the tracks are very large, others quite small, while some would seem, from the way in which the footsteps follow each other, to have been quadrupedal, and others bipedal. We can even measure the length of their strides, following the impressions which, from their succession in a continuous line, mark the walk of a single animal.* The fact that we find these footprints without any bones or other remains to indicate the animals by which they were made is accounted for by the mode of deposition of the sandstone. It is very unfavorable for the preservation of bones; but, being composed of minute sand mixed with mud, it affords an admirable substance for the reception of these impressions, which have been thus cast in a mould, as it were, and preserved through ages.

These animals must have been large, when full-grown, for we find strides measuring six feet between, evidently belonging to the same animal. In the quadrupedal tracks, the front seem to have

---

* For all details respecting these tracks see Hitchcock's *Ichnology of New England*. Boston, 1858. 4to.

been smaller than the hind ones. Some of the tracks show four toes all turned forward, while in others three toes are turned forward and one backward. It happened that the first tracks found belonged to the latter class; and they very naturally gave rise to the idea that these impressions were made by birds, on account of this formation of the foot. This, however, is a mere inference; and since the inductive method is the only true one in science, it seems to me that we should turn to the facts we have in our possession for the explanation of these mysterious footprints, rather than endeavor to supply by assumption those which we have not. As there are no bones found in connection with these tracks, the only way to arrive at their true character, in the present state of our knowledge, is by comparing them with bones found in other localities in the deposits of the same period in the world's history. Now there have never been found in the Trias any remains of Birds, while it contains innumerable bones of Reptiles; and therefore I think that we shall eventually find the solution of this mystery in the latter class.

It is true that the bones of the Triassic Reptiles are scattered and disconnected;* no complete skeleton has yet been discovered, nor has any foot

---

* See the Investigations of Hermann von Meyer on Triassic Reptiles.

been found; so that no direct comparison can be made with the steps. It is, however, my belief, from all we know of the character of the Animal Kingdom in those days, that these animals were reptilian, but combined, like so many of the early types, characters of their own class with those of higher animals yet to come. It seems to me probable, that, in those tracks where one toe is turned backward, the impression is made not by a toe, but by a heel, or by a long sole projecting backward; for it is not pointed, like those of the front toes, but is blunt. It is true that there is a division of joints in the toes, which seems in favor of the idea that they were those of Birds; for when the three toes are turned forward, there are two joints on the inner one, three on the middle, and four on the outer one, as in Birds. But this feature is not peculiar to Birds; it is found in Turtles also. The correspondence of these footprints with each other leaves no doubt that they were all by one kind of animal; for both the bipedal and the quadrupedal tracks have the same character. The only quadrupedal animals now known to us which walk on two legs are the Kangaroos. They raise themselves on their hind legs, using the front ones to bring their food to their mouth. They leap with the hind legs, sometimes bringing down their front feet to steady themselves after the spring, and making use also of their tails, to balance the body after leaping. In these tracks we find traces of a tail between the feet. I do not bring this forward as any evidence that these animals were allied to Kangaroos, since I believe that nothing is more injurious in science than assumptions which do not rest on a broad basis of facts; but I wish only to show that these tracks recall other animals besides Birds, with which they have been universally associated. And seeing, as we do, that so many of the early types prophesy future forms, it seems not improbable that they may have belonged to animals which combined with reptilian characters some birdlike features, and also some features of the earliest and lowest group of Mammalia, the Marsupials. To sum up my opinion respecting these footmarks, I believe that they were made by animals of a prophetic type, belonging to the class of Reptiles, and exhibiting many synthetic characters.

The more closely we study past creations, the more impressive and significant do the synthetic types, presenting features of the higher classes under the guise of the lower ones, become. They hold the promise of the future. As the opening overture of an opera contains all the musical elements to be therein developed, so this living prelude of the Creative work comprises all the organic elements to be successively developed in the course of time. When Cuvier first saw the teeth of a Wealden Reptile, he pronounced them to be those of a Rhinoceros, so mammalian were they in their appearance. So, when Sommering first saw the remains of a Jurassic Pterodactyl, he pronounced them to be those of a Bird. These mistakes were not due to a superficial judgment in men who knew Nature so well, but to this prophetic character in the early types themselves, in which features were united never known to exist together in our days, and presenting a kind of combination wholly new to scientific men at that time.

The Jurassic epoch, next in succession, was a very important one in the history of Europe. It completed the junction of several of the larger islands, filling the channel between the central plateau of France and the Belgian island, as well as that between the former and the island of Bretagne, so that France was now a sort of crescent of land holding a Jurassic sea in its centre, Bretagne and Belgium forming the two horns. This Jurassic basin or inland sea united England and France, and it may not be amiss to say a word here of its subsequent transformations. During the long succession of Jurassic periods, the deposits of that epoch, chiefly limestone and clays, with here and there a bed of sand, were accumulated at its bottom. Upon these followed the chalk deposits of the Cretaceous epoch, until the basin was gradually filled, and partially, at least, turned to dry land. But at the close of the Cretaceous epoch a fissure was formed, allowing the entrance of the sea at the western end, so that the constant washing of the tides and storms wore away the lower, softer deposits, leaving the overhanging chalk cliffs unsupported. These latter, as their supports were undermined, crumbled down, thus widening the channel gradually. This process must, of course, have gone on more rapidly at the western end, where the sea rushed in with most force, till the channel was worn through to the German Ocean on the other side, and the sea then began to act with like power at both ends of the channel. This explains its form, wider at the western end, narrower between Dover and Calais, and widening again at the eastern extremity. This ancient basis, extending from the centre of France into England, is rich in the remains of a number of successive epochs. Around its margin we find the Jurassic deposits, showing that there must have been some changes of level which raised the shores and prevented later accumulations from covering them, while in the centre the Jurassic deposits are concealed by those of the Cretaceous epoch above them, these being also partially hidden under the later Tertiary beds. Let us see, then, what this inland sea has to tell us of the organic world in the Jurassic epoch.

At that time the region where Lyme-Regis is now situated in modern England was an estuary on the shore of that ancient sea. About fifty years ago a discovery of large and curious bones, belonging to some animal unknown to the scientific world, turned the attention of naturalists to this locality, and since then such a quantity and variety of such remains have been found in that neighborhood as to show that the Sharks, Whales, Porpoises, etc., of the present ocean are not more numerous and diversified than were the inhabitants of this old bay or inlet. Among these animals, the Ichthyosauri (Fish-Lizards) form one of the best-known and most prominent groups. They are chiefly found in the Lias, the lowest set of beds of the Jurassic deposits, and seem to have come in with the close of the Triassic epoch. It is greatly to be regretted that whatever is known of the Triassic Reptiles antecedent to the Ichthyosauri still remains in the form of original papers, and is not yet embodied in text-books. They are quite as interesting, as curious, and as diversified as those of the Jurassic epoch, which are, however, much more extensively known, on account of the large collections of these animals belonging to the British Museum. It will be more easy to understand the structural relations of the latter, and their true position in the Animal Kingdom, when those which preceded them are better

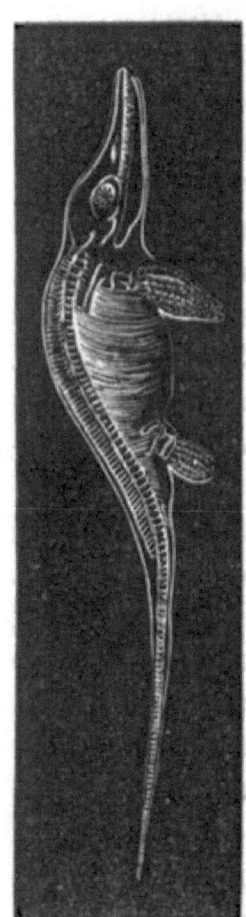

Fig. 1.

understood. One of the most remarkable and numerous of these Triassic Reptiles seems to have been an animal called Labyrinthodon, and resembling, in the form of the head, and in the two articulating surfaces at the juncture of the head with the backbone, the Frogs and Salamanders, though its teeth are like those of a Crocodile. As yet nothing has been found of these animals except the head, — neither the backbone nor the limbs; so that little is known of their general structure.

The Ichthyosauri (Figure 1) must have been very large, seven or eight feet being the ordinary length, while specimens measuring from twenty to thirty feet are not uncommon. The large head is pointed, like that of the Porpoise; the jaws contain a number of conical teeth, of reptilian form and character; the eyeball was very large, as may be seen by the socket, and it was supported by pieces of bone, such as we find now only in the eyes of birds of prey and in the bony fishes. The ribs begin at the neck and continue to the tail, and there is no distinction between head and neck, as in most Reptiles, but a continuous outline, as in Fishes. They had four limbs, not divided into fingers, but forming mere paddles. Yet fingers seem to be hinted at in these paddles, though not developed, for the bones are in parallel rows, as if to mark what might be such a division. The backbones are short, but very high, and the surfaces of articulation are hollow, conical cavities, as in Fishes, instead of ball-and-socket joints, as in Reptiles. The ribs are more complicated than in Vertebrates generally: they consist of several pieces, and the breast-bone is formed of a number of bones, making together quite an intricate bony network. There is only one living animal, the Crocodile, characterized by this peculiar structure of the breast-bone. The Ichthyosaurus is, indeed, one of the most remarkable of the synthetic types: by the shape of its head one would associate it with the Porpoises, while by its paddles and its long tail it reminds one of the whole group of Cetaceans to which the Porpoises belong; by its crocodilian teeth, its ribs, and its breast-bone, it seems allied to Reptiles; and by its uniform neck, not distinguished from the body, and the structure of the backbone, it recalls the Fishes.

Another most curious member of this group is

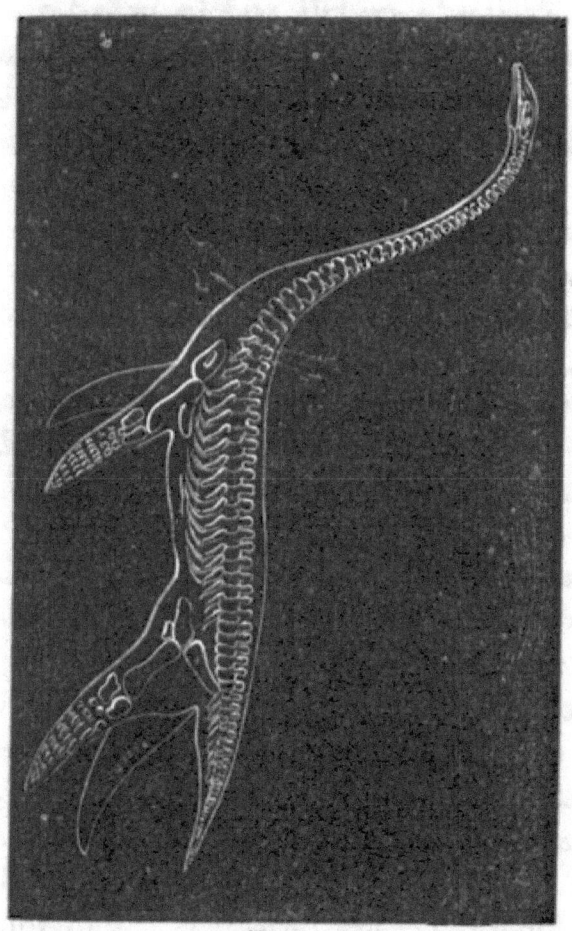

Fig. 2.

the Plesiosaurus, old Saurian (Figure 2). By its disproportionately long and flexible neck, and its

small, flat head, it unquestionably foreshadows
the Serpents, while by the structure of the backbone, the limbs, and the tail, it is closely allied
with the Ichthyosaurus. Its flappers are, however,
more slender, less clumsy, and were, no doubt,
adapted to more rapid motion than the fins of the
Ichthyosaurus, while its tail is shorter in proportion to the whole length of the animal. It seems
probable, from its general structure, that the
Ichthyosaurus moved like a Fish, chiefly by the
flapping of the tail, aided by the fins, while in the
Plesiosaurus the tail must have been much less
efficient as a locomotive organ, and the long,
snake-like, flexible neck no doubt rendered the
whole body more agile and rapid in its movements. In comparing the two, it may be said,
that, as a whole, the Ichthyosaurus, though belonging by its structure to the class of Reptiles,
has a closer external resemblance to the Fishes,
while the Plesiosaurus is more decidedly reptilian
in character. If there exists any animal in our
waters, not yet known to naturalists, answering
to the descriptions of the "Sea-Serpent," it must
be closely allied to the Plesiosaurus. The occurrence in the fresh waters of North America of a
Fish, the Lepidosteus, which is closely allied to
the fossil Fishes found with the Plesiosaurus in
the Jurassic beds, renders such a supposition
probable

Of all these strange old forms, so singularly uniting features of Fishes and Reptiles, none has given rise to more discusion than the Pterodactylus, (Figure 3,) another of the Saurian tribe,

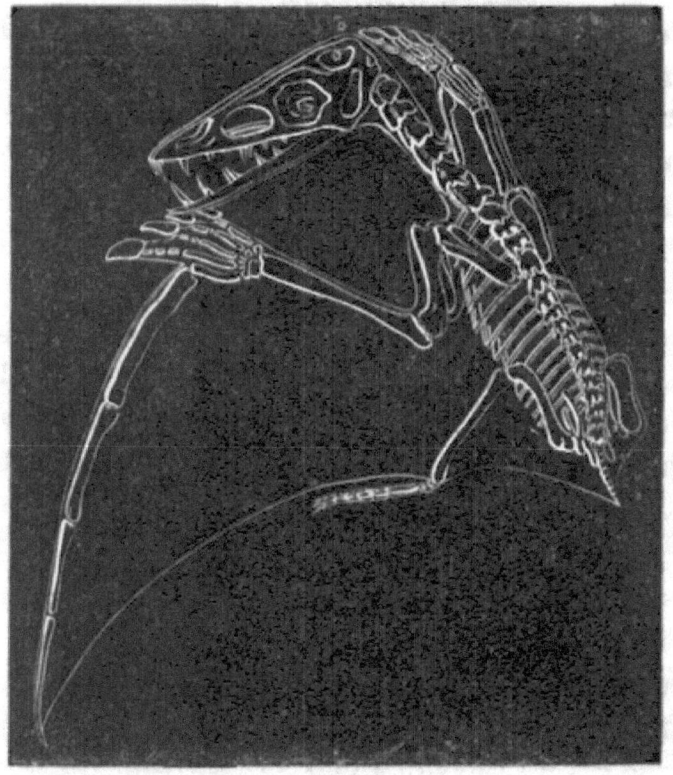

Fig. 3.

associated, however, with Birds by some naturalists, on account of its large wing-like appendages. From the extraordinary length of its anterior limbs, they have generally been described as wings, and the animal is usually represented as a flying Reptile. But if we consider its whole structure, this does not seem probable, and I believe it to have been an essentially aquatic animal, moving after the fashion of the Sea-Turtle. Its so-called wings resemble in structure the front paddles of the Sea-Turtles far more than the wings of a Bird; differing from them, indeed, only by the extraordinary length of the inner toe, while the outer ones are comparatively much shorter. But, notwithstanding this difference, the hand of the Pterodactylus is constructed like that of an aquatic swimming marine Reptile; and I believe, that, if we represent it with its long neck stretched upon the water, its large head furnished with powerful, well-armed jaws, ready to dive after the innumerable smaller animals living in the same ocean, we shall have a more natural picture of its habits than if we consider it as a flying animal, which it is generally supposed to have been. It has not the powerful breast-bone, with the large projecting keel along the middle line, such as exists in all the flying animals. Its breast-bone, on the contrary, is thin and flat, like that of the present Sea-Turtle; and if it moved through the water by the help of its long flappers, as the Sea-Turtle does now, it could well dispense with that powerful construction of the breast-bone so essential to all animals which fly through the air. Again, the powerful teeth, long and conical, placed at considerable intervals in the jaw, constitute a feature common to all predaceous aquatic animals, and would seem to have been utterly useless in a flying animal at that time, since there were no aërial beings of any size to prey upon. The Dragon-Flies found in the same deposits with the Pterodactylus were certainly not a game requiring so powerful a battery of attack.

The Fishes of the Jurassic sea were exceedingly numerous, but were all of the Ganoid and Selachian tribes. It would weary the reader, were I to introduce here any detailed description of them, but they were as numerous and varied as those living in our present waters. There was the Hybodus, with the marked furrows on the spines and the strong hooks along their margin, — the huge Chimera, with its long whip, its curved bone over the back, and its parrot-like bill, — the Lepidotus, with its large square scales, its large head, its numerous rows of teeth, one within another, forming a powerful grinding apparatus, — the Microdon, with its round, flat body, its jaw paved with small grinding teeth, — the swift Aspidorhynchus, with its long, slender body and massive tail, enabling it to strike the water powerfully and dart forward with great rapidity. There were also a host of small Fishes, comparing with those above mentioned as our

Perch, Herring, Smelts, etc., compare with our larger Fishes; but, whatever their size or form, all the Fishes of those days had the same hard scales fitting to each other by hooks, instead of the thin membranous scales overlapping each other at the edge, like the common Fishes of more modern times. The smaller Fishes, no doubt, afforded food to the larger ones, and to the aquatic Reptiles. Indeed, in parts of the intestines of the Ichthyosauri, and in their petrified excrements, have been found the scales and teeth of these smaller Fishes perfectly preserved. It is amazing that we can learn so much of the habits of life of these past creatures, and know even what was the food of animals existing countless ages before man was created.

There are traces of Mammalia in the Jurassic deposits, but they were of those inferior kinds known now as Marsupials, and no complete specimens have yet been found.

The Articulates were largely represented in this epoch. There were already in the vegetation a number of Gymnosperms, affording more favorable nourishment for Insects than the forests of earlier times; and we accordingly find that class in larger numbers than ever before, though still meagre in comparison with its present representation. Crustacea were numerous, — those of the Shrimp and Lobster kinds prevailing, though in some of the Lobsters we have the first advance towards the highest class of Crustacea in the expansion of the transverse diameter now so characteristic of the Crabs. Among Mollusks we have a host of gigantic Ammonites; and the naked Cephalopods, which were in later times to become the prominent representatives of that class, already begin to make their appearance. Among Radiates, some of the higher kinds of Echinoderms, the Ophiurans, and Echinoids, take the place of the Crinoids, and the Acalephian Corals give way to the Astræan and Meandrina-like types, resembling the Reef-Builders of the present time.

I have spoken especially of the inhabitants of the Jurassic sea lying between England and France, because it was there that were first found the remains of some of the most remarkable and largest Jurassic animals. But wherever these deposits have been investigated, the remains contained in them reveal the same organic character, though, of course, we find the land Reptiles only where there happen to have been marshes, the aquatic Saurians wherever large estuaries or bays gave them an opportunity of coming in near shore, so that their bones were preserved in the accumulations of mud or clay constantly collecting in such localities,— the Crustacea, Shells, or Sea-Urchins on the old sea-beaches, the Corals in the neighborhood of coral reefs, and so on. In short, the distribution of animals then as now was in accordance with their nature and habits, and we shall seek vainly for them in the localities where they did not belong.

But when I say that the character of the Jurassic animals is the same, I mean, that, wherever a Jurassic sea-shore occurs, be it in France, Germany, England, or elsewhere throughout the world, the Shells, Crustacea, or other animals found upon it have a special character, and are not to be confounded by any one thoroughly acquainted with these fossils with the Shells or Crustacea of any preceding or subsequent time, — that, where a Jurassic marsh exists, the land Reptiles inhabiting it are Jurassic, and neither Triassic nor Cretaceous, — that a Jurassic coral reef is built of Corals belonging as distinctly to the Jurassic creation as the Corals on the Florida reefs belong to the present creation, — that, where some Jurassic bay or inlet is disclosed to us with the Fishes anciently inhabiting it, they are as characteristic of their time as are the Fishes of Massachusetts Bay now.

And not only so, but, while this unity of creation prevails throughout the entire epoch as a whole, there is the same variety of geographical distribution, the same circumscription of faunæ within distinct zoölogical provinces, as at the present time. The Fishes of Massachusetts Bay are not the same as those of Chesapeake Bay, nor those of Chesapeake Bay the same as those of Pamlico Sound, nor those of Pamlico Sound the same as those of the Florida coast. This division of the surface of the earth into given areas within which certain combinations of animals and plants are confined is not peculiar to the present creation, but has prevailed in all times, though with ever-increasing diversity, as the surface of the earth itself assumed a greater variety of climatic conditions. D'Orbigny and others were mistaken in assuming that faunal differences have been introduced only in the last geological epochs. Besides these adjoining zoölogical faunæ, each epoch is divided, as we have seen, into a number of periods, occupying successive levels one above another, and differing specifically from each other in time as zoölogical provinces differ from each other in space. In short, every epoch is to be looked upon from two points of view: as a unit, complete in itself, having one character throughout, and as a stage in the progressive history of the world, forming part of an organic whole.

As the Jurassic epoch was ushered in by the upheaval of the Jura, so its close was marked by the upheaval of that system of mountains called the Côte d'Or. With this latter upheaval began the Cretaceous epoch, which we will examine with special reference to its subdivision into periods, since the periods in this epoch have been clearly distinguished, and investigated with especial care. I have alluded in the preceding article to the immediate contact of the Jurassic and Cretaceous epochs in Switzerland, affording peculiar facilities for the direct comparison of their organic remains. But the Cretaceous deposits are well known, not only in this inland sea of ancient Switzerland, but in a number of European basins, in France, in the Pyrenees, on the Mediterranean shores, and also in Syria, Egypt, India, and Southern Africa, as well as on our own continent. In all these localities, the Cretaceous remains, like those of the Jurassic epoch, have one organic character, distinct and unique. This fact is especially significant, because the contact of their respective deposits is in many localities so immediate and continuous that it affords an admirable test for the development-theory. If this is the true mode of origin of animals, those of the later Jurassic beds must be the progenitors of those of the earlier Cretaceous deposits. Let us see now how far this agrees with our knowledge of the physiological laws of development.

Take first the class of Fishes. We have seen that in the Jurassic periods there were none of our common Fishes, none corresponding to our Herring, Pickerel, Mackerel, and the like,— no Fishes, in short, with thin membranous scales, but that the class was represented exclusively by those with hard, flint-like scales. In the Cretaceous epoch, however, we come suddenly upon a horde of Fishes corresponding to our smaller common Fishes of the Pickerel and Herring tribes, but principally of the kinds found now in tropical waters; there are none like our Cods, Haddocks, etc., such as are found at present in the colder seas. The Fishes of the Jurassic epoch corresponding to our Sharks and Skates and Gar-Pikes still exist, but in much smaller proportion, while these more modern kinds are very numerous. Indeed, a classification of the Cretaceous Fishes would correspond very nearly to one founded on those now living. Shall we, then, suppose that the large reptilian Fishes of the Jurassic time began suddenly to lay numerous broods of these smaller, more modern, scaly Fishes? And shall we account for the diminution of the previous forms by supposing that in order to give a fair chance to the new kinds they brought them forth in large numbers, while they reproduced their own kind less abundantly? According to very careful estimates, if we accept this view, the progeny of the Jurassic Fishes must have borne a proportion of about ninety per cent of entirely new types to some ten per cent of those resembling the parents. One would like a fact or two on which to rest so very extraordinary a reversal of all known physiological laws of reproduction, but, unhappily, there is not one.

Still more unaccountable, upon any theory of development according to ordinary laws of reproduction, are those unique, isolated types limited to a single epoch, or sometimes even to a single period. There are some very remarkable instances of this in the Cretaceous deposits. To make my statement clearer, I will say a word of the sequence of these deposits and their division into periods.

These Cretaceous beds were at first divided only into three sets, called the Neocomian, or lower deposits, the Green-Sands, or middle deposits, and the Chalk, or upper deposits. The Neocomian, the lower division, was afterwards subdivided into three sets of beds, called the Lower, Middle, and Upper Neocomian by some geologists, the Valengian, Neocomian, and Urgonian by others. These three periods are not only traced in immediate succession, one above another, in the transverse cut before described, across the mountain of Chaumont, near Neufchâtel, but they are also traced almost on one level along the plain at the foot of the Jura. It is evident that by some disturbance of the surface the eastern end of the range was raised slightly, lifting the lower or Valengian deposits out of the water, so that they remain uncovered, and the next set of deposits, the Neocomian, is accumulated along their base, while these in their turn are slightly raised, and the Urgonian beds are accumulated against them a little lower down. They follow each other from east to west in a narrower area, just as the Azoic, Silurian, and Devonian deposits follow each other from north to south in the northern part of the United States. The Cretaceous deposits have been intimately studied in various localities by different geologists, and are now subdivided into at least ten, or it may be fifteen or sixteen distinct periods, as they stand at present. This is, however, but the beginning of the work; and the recent investigations of the French geologist, Coquand, indicate that several of these periods at least are susceptible of further subdivision. I present here a table enumerating the periods of the Cretaceous epoch best known at present, in their sequence, because I want to show how sharply and in how arbitrary a manner, if I may so express it, new forms are introduced. The names are simply derived from the localities, or from some circumstances connected with the locality where each period has been studied.

THE GEOLOGICAL MIDDLE AGE.

Table of Periods in the Cretaceous Epoch.

Maestrichtian	⎫ Chalk.
Senonian	⎭
Turonian	⎫ Chalk Marl.
Cenomanian	⎭
Albian	⎫
Aptian	⎬ Green Sands.
Rhodanian	⎭
Urgonian	⎫
Neocomian	⎬ Wealden.
Valengian	⎭

One of the most peculiar and distinct of those unique types alluded to above is that of the Rudistes, a singular Bivalve, in which the lower valve is very deep and conical, while the upper valve sets into it as into a cup. The subjoined wood-cut represents such a Bivalve. These Rudistes are found suddenly in the Urgonian deposits; there are none in the two preceding sets of beds; they disappear in the three following periods, and reappear again in great numbers in the Cenomanian, Turonian, and Senonian periods, and disappear again in the succeeding one. These can hardly 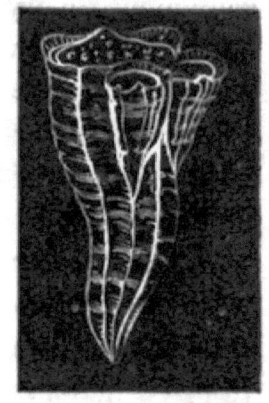 be missed from any negligence or oversight in the examination of these deposits, for they are by no means rare. They are found always in great numbers, occupying crowded beds, like Oysters in the present time. So numerous are they, where they occur at all, that the deposits containing them are called by many naturalists the first, second, third, and fourth *bank* of Rudistes. Which of the ordinary Bivalves, then, gave rise to this very remarkable form in the class, allowed it to die out, and revived it again at various intervals? This is by no means the only instance of the same kind. There are a number of types making their appearance suddenly, lasting during one period or during a succession of periods, and then disappearing forever, while others, like the Rudistes, come in, vanish, and reappear at a later time.

I am well aware that the advocates of the development-theory do not state their views as I have here presented them. On the contrary, they protest against any idea of sudden, violent, abrupt changes, and maintain that by slow and imperceptible modifications during immense periods of time these new types have been introduced without involving any infringement of the ordinary processes of development; and they account for the entire absence of corroborative facts in the past history of animals by what they call the "imperfection of the geological record." Now, while I admit that our knowledge of geology is still very incomplete, I assert that just where the direct sequence of geological deposits is needed for this evidence, we have it. The Jurassic beds, without a single modern scaly Fish, are in immediate contact with the Cretaceous beds, in which the Fishes of that kind are proportionately almost as numerous as they are now; and between these two sets of deposits there is not a trace of any transition or intermediate form to unite the reptilian Fishes of the Jurassic with the common Fishes of the Cretaceous times. Again, the Cretaceous beds in which the crowded banks of Rudistes, so singular and unique in form, first make their appearance, follow immediately upon those in which all the Bivalves are of an entirely different character. In short, the deposits of this year along any sea-coast or at the mouth of any of our rivers do not follow more directly upon those of last year than do these successive sets of beds of past ages follow upon each other. In making these statements, I do not forget the immense length of the geological periods; on the contrary, I fully accede to it, and believe that it is more likely to have been underrated than overstated. But let it be increased a thousand-fold, the fact remains, that these new types occur commonly at the dividing line where one period joins the next, just on the margin of both.

For years I have collected daily among some

of these deposits, and I know the Sea-Urchins, Corals, Fishes, Crustacea, and Shells of those old shores as well as I know those of Nahant Beach, and there is nothing more striking to a naturalist than the sudden, abrupt changes of species in passing from one to another. In the second set of Cretaceous beds, the Neocomian, there is found a little Terebratula (a small Bivalve Shell) in immense quantities: they may actually be collected by the bushel. Pass to the Urgonian beds, resting directly upon the Neocomian, and there is not one to be found, and an entirely new species comes in. There is a peculiar Spatangus (Sea-Urchin) found throughout the whole series of beds in which this Terebratula occurs. At the same moment that you miss the Shell, the Sea-Urchin disappears also, and another takes its place. Now, admitting for a moment that the later can have grown out of the earlier forms, I maintain, that, if this be so, the change is immediate, sudden, without any gradual transitions, and is, therefore, wholly inconsistent with all our known physiological laws, as well as with the transmutation-theory.

There is a very singular group of Ammonites in the Cretaceous epoch, which, were it not for the suddenness of its appearance, might seem rather to favor the development-theory, from its great variety of closely allied forms. We have traced the Chambered Shells from the straight, simple ones of the earliest epochs up to the intricate and closely coiled forms of the Jurassic epoch. In the so-called Portland stone, belonging to the upper set of Jurassic beds, there is only one type of Ammonite; but in the Cretaceous beds, immediately above it, there set in a number of different genera and distinct species, including the most fantastic and seemingly abnormal forms. It is as if the close coil by which these shells had been characterized during the Middle Age had been suddenly broken up and decomposed into an endless variety of outlines. Some of these new types still retain the coil, but the whorls are much less compact than before, as in the Crioceras (Figure 1); in others, the di-

Fig. 1.

Fig. 2.

rection of the coil is so changed as to make a spiral, as in the Turrilites (Figure 2); or the shell starts with a coil, then proceeds in a straight line, and changes to a curve again at the other extremity, as in the Ancyloceras (Figure 3), or in the Scaphites (Figure 4), in which the first coil is somewhat closer than in the Ancyloceras; or the tendency to a coil is reduced to a single curve, so as to give the shell the outline of a

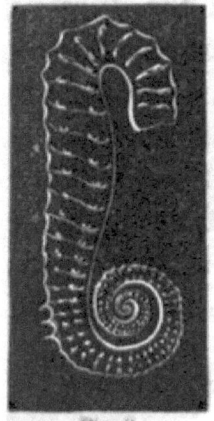
Fig. 3.

Fig. 4.

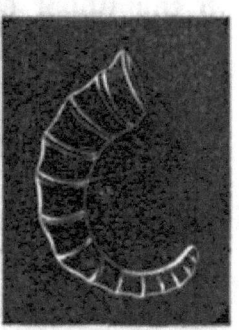
Fig. 5.

horn, as in the Toxoceras (Figure 5); or the coil is entirely lost, and the shell reduced to its primitive straight form, as in the Baculites

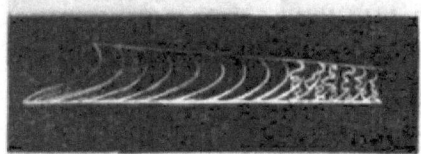
Fig. 6.

(Figure 6), which, except for their undulating partitions, might be mistaken for the Orthoceratites of the Silurian and Devonian epochs. I have presented here but a few species of these extraordinary Cretaceous Ammonites, and, strange to say, with this breaking-up of the type into a number of fantastic and often contorted shapes, it disappears. It is singular that forms so unusual and so contrary to the previous regularity of this group should accompany its last stage of existence, and seem to shadow forth by their strange contortions the final dissolution of their type. When I look upon a collection of these old shells, I can never divest myself of an impression that the contortions of a death-struggle have been made the pattern of living types, and with that the whole group has ended.

Now shall we infer that the compact, closely coiled Ammonites of the Jurassic deposits, while continuing their own kind, brought forth a variety of other kinds, and so distributed these new organic elements as to produce a large number of distinct genera and species? I confess that these ideas are so contrary to all I have learned from Nature in the course of a long life that I should be forced to renounce completely the results of my studies in Embryology and Palæontology before I could adopt these new views of the origin of species. And while the distinguished originator of this theory is entitled to our highest respect for his scientific researches, yet it should not be forgotten that the most conclusive evidence brought forward by him and his adherents is of a negative character, drawn from a science in which they do not pretend to have made personal investigations, that of Geology, while the proofs they offer us from their own departments of science, those of Zoölogy and Botany, are derived from observations, still very incomplete, upon domesticated animals and cultivated plants, which can never be made a test of the origin of wild species.*

In my next article I shall show the relation between the Cretaceous and Tertiary epochs, and see whether there is any reason to believe that the gigantic Mammalia of more modern times were derived from the Reptiles of the Secondary age.

---

* The advocates of the development-theory allude to the metamorphosis of animals and plants as supporting their view of a change of one species into another. They compare the passage of a common leaf into the calyx or crown-leaves in plants, or that of a larva into a perfect insect, to the passage of one species into another. The only objection to this argument seems to be, that, whereas Nature daily presents us myriads of examples of the one set of phenomena, showing it to be a norm, not a single instance of the other has ever been known to occur either in the animal or in the vegetable kingdom.

VII.

THE TERTIARY AGE, AND ITS CHARACTERISTIC ANIMALS.

IN entering upon the Tertiaries, we reach that geological age which, next to his own, has the deepest interest for man. The more striking scenes of animal life, hitherto confined chiefly to the ocean, are now on land; the extensive sheets of fresh water are filled with fishes of a comparatively modern character,—with Whitefish, Pickerel, Perch, Eels, etc.,—while the larger quadrupeds are introduced upon the continents so gradually prepared to receive them. The connection of events throughout the Tertiaries, considered as leading up to the coming of man, may be traced not only in the physical condition of the earth, and in the presence of the large terrestrial Mammalia, but also in the appearance of those groups of animals and plants which we naturally associate with the domestic and social existence of man. Cattle and Horses are first found in the middle Tertiaries; the grains, the Rosaceæ, with their variety of fruits, the tropical fruit-trees,

Oranges, Bananas, etc., the shade- and cluster-trees, so important to the comfort and shelter of man, are added to the vegetable world during these epochs. The fossil vegetation of the Tertiaries is, indeed, most interesting from this point of view, showing the gradual maturing and completion of those conditions most intimately associated with human life. The earth had already its seasons, its spring and summer, its autumn and winter, its seed-time and harvest, though neither sower nor reaper was there; the forests then, as now, dropped their thick carpet of leaves upon the ground in the autumn, and in many localities they remain where they originally fell, with a layer of soil between the successive layers of leaves, — a leafy chronology, as it were, by which we read the passage of the years which divided these deposits from each other. Where the leaves have fallen singly on a clayey soil favorable for receiving such impressions, they have daguerrotyped themselves with the most wonderful accuracy, and the Oaks, Poplars, Willows, Maples, Walnuts, Gum- and Cinnamon-trees, etc., of the Tertiaries are as well known to us as are those of our own time.

It was an eventful day, not only for science, but for the world, when a Siberian fisherman chanced to observe a singular mound lying near the mouth of the River Lena, where it empties into the Arctic Ocean. During the warmer summer-weather, he noticed, that, as the snow gradually melted, this mound assumed a more distinct and prominent outline, and at length, on one side of it, where the heat of the sun was greatest, a dark body became exposed, which, when completely uncovered, proved to be that of an immense elephant, in so perfect a state of preservation that the dogs and wolves were attracted to it as by the smell of fresh meat, and came to feed upon it at night. The man knew little of the value of his discovery, but the story went abroad, and an Englishman travelling in Russia, being curious to verify it, visited the spot, and actually found the remains where they had been reported to lie, on the frozen shore of the Arctic Sea, — strange burial-place enough for an animal never known to exist out of tropical climates. Little beside the skeleton was left, though parts of the skin remained covered with hair, showing how perfect must have been the condition of the body when first exposed. The tusks had been sold by the fisherman; but Mr. Adams succeeded in recovering them; and collecting all the bones, except those of one foot, which had been carried off by the wolves, he had them removed to St. Petersburg, where the skeleton now stands in the Imperial Museum. The inhabitants of Siberia seem to be familiar with this animal, which they designate by the name of *Mammoth*, while naturalists call it *Elephas primigenius*. The circumstance that they abound in the frozen drift of the great northern plain of Asia, and are occasionally exposed in consequence of the wearing of the large rivers traversing Siberia, has led to the superstition among the Tongouses, that the Mammoths live under ground, and die whenever, on coming to the surface, the sunlight falls upon them.

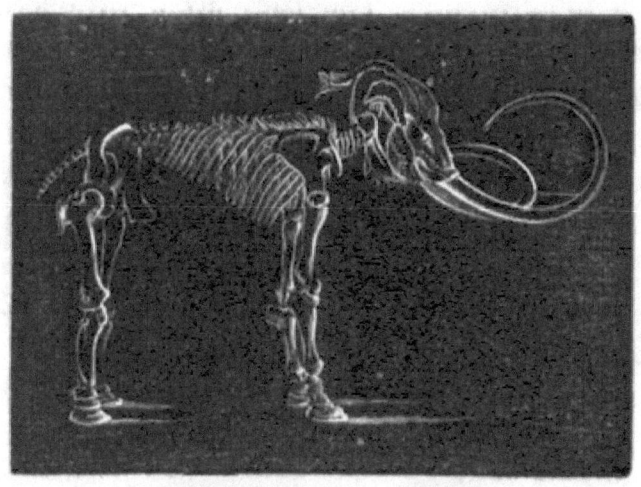

Mammoth, the Elephas primigenius.

Had this been the only creature of the kind found so far from the countries to which elephants are now exclusively confined, it might have been believed that some strange accident had brought it to the spot where it was buried. But it was not long before similar remains were found in various parts of Europe, — in Russia, in Germany, in Spain, and in Italy. The latter were readily accounted for by the theory that they must be the remains of the Carthaginian elephants brought over by the armies of Hannibal, while it was suggested that the others might have been swept from India by some great flood, and stranded where they were found. It was Cuvier, entitled by his intimate acquaintance with the anatomy of living animals to an authoritative opinion in such matters, who first dared to assert that these remains belonged to no elephant of our period. He rested this belief upon structural evidence, and insisted that an Indian elephant, brought upon the waves of a flood to Siberia, would be an Indian elephant still, while all these remains differed in structure from any species existing at present. This statement aroused research in every direction, and the number of fossil Mammalia found within the next few years, and proved by comparison to be different from any living species, soon demonstrated the truth of his conclusion.

Shortly after the discovery of fossil elephants had opened this new path of investigation, some curious bones were found by some workmen in the quarries of Montmartre, near Paris, and brought to Cuvier for examination. Although few in numbers, and affording but very scanty *data* for such a decision, he at once pronounced them to be the remains of some extinct animal preceding the present geological age. Here, then, at his very door, as it were, was a settlement of that old creation in which he could pursue the inquiry, already become so important in its bearings. It was not long before other bones of the same kind were found, though nothing as yet approaching an entire skeleton. However, with such means as he had, Cuvier began a comparison with all the living Mammalia, — with the human skeleton first, with Monkeys, with the larger Carnivora and Ruminants, then with all the smaller Mammalia, then with the Pachyderms; and here, for the first time, he began to find some resemblance. He satisfied himself that the animal must have belonged to the family of Pachyderms; and he then proceeded to analyze and compare all the living species, till he had collected ample evidence to show that the bones in question did not correspond with any species, and could not even be referred to any genus, now in existence. At length there was discovered at Montmartre an upper jaw of the same animal, — next a lower jaw, matching the upper one, and presently a whole head with a few backbones was brought to light. These were enough, with Cuvier's vast knowledge of animal structure, to give him a key to the whole skeleton. At about the same time, in the same locality, were found other bones and teeth also, differing from those first discovered, and yet equally unlike those of any living animal. The first evidently belonging to some stout and heavy animal, the others were more slender and of lighter build. From these fragments, ample evidence to him of his results, he drew the outlines of two animals: one which he called the Palæotherium (old

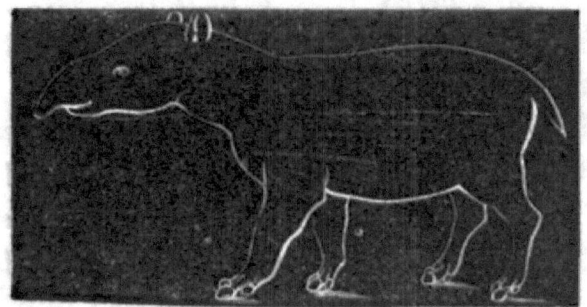

Palæotherium.

animal), a figure of which is given in the above wood-cut, and the other Anoplotherium (animal without fangs). He presented these figures with an explanatory memoir at the Academy, and announced them as belonging to some creation preceding the present, since no such animals had ever existed in our own geological period. Such a statement was a revelation to the scientific world: some looked upon it with suspicion and distrust; others, who knew more of comparative anatomy, hailed it as introducing a new era in science; but it was not till complete specimens were actually found of animals corresponding perfectly to those figured and described by Cuvier, and proving beyond a doubt their actual existence in ancient times, that all united in wonder and admiration at the result obtained by him with such scanty means.

It would seem that the family of Pachyderm was largely represented among the early Mammalia; for, since Cuvier named these species, a number of closely allied forms have been found in deposits belonging to the same epoch. Of course, the complete specimens are rare; but the fragments of such skeletons occur in abundance, showing that these old-world Pachyderms, resembling the Tapirs more than any other living representatives of the family, were very numerous in the lower Tertiaries.

There is, however, one animal now in existence, forming one of those singular links before alluded to between the present and the past, of which I will say a few words here, though its relation is rather with a later group of Tertiary Pachyderms than with those described by Cuvier. On the coast of Florida there is an animal of very massive, clumsy build, long considered to be a Cetacean, but now recognized, by some naturalists at least, as belonging to the order of Pachyderms. In form it resembles the Cetaceans, though it has a fan-shaped tail, instead of the broad flapper of the Whales. It inhabits fresh waters or shoal waters, and is not so exclusively aquatic as the oceanic Cetaceans. Its most striking feature is the form of the lower jaw, which is bent downward, with the front teeth hanging from it. This animal is called the Manatee, or Sea-Cow. There are three species known to naturalists,— one in Tampa Bay, one in the Amazon, and one in the Red Sea. In the Tertiary deposits of Germany there has been found an animal allied in some of its features to those described by Cuvier, but it has the crown of its teeth folded like the Tapir, while the lower jaw is turned down with a long tusk growing from it. This animal has been called the Dinotherium. A part of the head, showing the heavy jaws and the formidable tusk, is represented in the subjoined wood-cut. 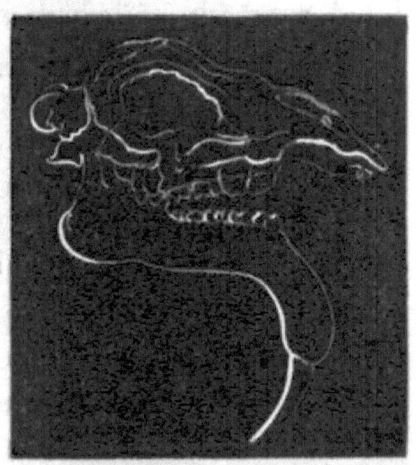 Its hanging lower jaw, with the protruding tusk, corresponds perfectly to the formation of the lower jaw and teeth in the Manatee. Some resemblance of the Dinotherium to the Mastodon suggested a comparison with that animal as the next step in the investigation, when it was found that at the edge of the lower jaw of the latter there was a pit with a small projecting tooth, also corresponding exactly in its position to the tusk in the Dinotherium. The Elephant was now examined; and in him also a rudimentary tooth appeared in the lower jaw, not cut through, but placed in the same relation to the jaw and the other teeth as that of the Mastodon. It would seem, then, that the Manatee makes one in this series of Dinotherium, Mastodon, and Elephant, and represents the aquatic Pachyderms, occupying the same relation to the terrestrial Pachyderms as the Seals bear to the terrestrial Carnivora, and, like them, lowest in structure among their kind.

The announcement of Cuvier's results stimulated research, and from this time forward Tertiary Mammalia became the subject of extensive and most important investigations among naturalists. The attention of collectors once drawn to these remains, they were found in such numbers that the wonder was how they had been so long hidden from the observation of men. They remind us chiefly of tropical animals; indeed, Tigers, Hyenas, Rhinoceroses, Hippopotamuses, Mastodons, and Elephants had their home in countries which now belong to the Cold Temperate Zone, showing that the climate in these latitudes was much milder then than it is at present. Bones of many of these animals were found in caverns in Germany, France, Italy, and England. Perhaps the story of Kirkdale Cave, where the first important discovery of this kind was made on English soil, may not be so well known to American readers as to forbid its repetition here.

It was in the summer of 1821 that some workmen, employed in quarrying stone upon the slope of a limestone hill at Kirkdale, in Yorkshire, came accidentally upon the mouth of a cavern. Overgrown with grass and bushes, the mouth of this cave in the hill-side had been effectually closed against all intruders, and it was strange that its existence had never been suspected. The hole was small, but large enough to admit a man on his hands and knees; and the workmen, creeping in through the opening, found that it led into a cavern, broad in some parts, but low throughout. There were only a few spots where a man could stand upright; but it was quite extensive, with branches opening out from it, some of which have not yet been explored. The whole floor was strewn, from one end to the other, with hundreds of bones, like a huge dog-kennel. The workmen wondered a little at their discovery, but, remembering that there had been a murrain among the cattle in this region some years before, they came to the conclusion that these must be the bones of cattle that had died in great numbers at that time; and, having so settled the matter to their own satisfaction, they took little heed to the bones, but threw many of them out on the road with the common limestone. Fortunately, a gentleman, living in the neighborhood, whose attention had been attracted to them, preserved them from destruction; and a few months after the discovery of the cave, Dr. Buckland, the great English geologist, visited Kirkdale, to examine its strange contents, which proved indeed stranger than any one had imagined; for many of these remains belonged to animals never before found in England. The bones of Hyenas, Tigers, Elephants, Rhinoceroses, and Hippopotamuses were mingled with those of Deer, Bears, Wolves, Foxes, and many smaller creatures. The bones were gnawed, and many were broken, evidently not by natural decay, but seemed to have been snapped violently apart. After the most complete investigation of the circumstances, Dr. Buckland convinced himself, and proved to the satisfaction of all scientific men, that the cave had been a den of Hyenas* at a time when they, as well as

* Among the other facts showing that Kirkdale Cave had been the den of these animals, and not tenanted as their home by any of the other creatures whose remains occurred there, were the excrements of the Hyenas found in considerable quantity by Dr. Buckland, and identified as such by the keeper of a menage-

Tigers, Elephants, Rhinoceroses, etc., existed in England in as great numbers as they now do in the wildest parts of tropical Asia or Africa. The narrow entrance to the cave still retains the marks of grease and hair, such as one may see on the bars of a cage in a menagerie against which the imprisoned animals have been in the habit of rubbing themselves constantly, and there were marks of the same kind on the floor and walls. It was evident that the Hyenas were the lords of this ancient cavern, and the other animals their unwilling guests; for the remains of the latter were those which had been most gnawed, broken, and mangled; and the head of an enormous Hyena, with gigantic fangs found complete, bore ample evidence to their great size and power. Some of the animals, such as the Elephants, Rhinoceroses, etc., could not have been brought into the cave without being first killed and torn to pieces, for it is not large enough to admit them. But their gnawed and broken bones attest, nevertheless, that they were devoured like the rest; and probably the Hyenas then had the same propensity which characterizes those of our own

rie. Any one who may wish to read the whole history of Dr. Buckland's investigations of this matter, showing the patience and sagacity with which he collected and arranged the evidence, will find a full account of Kirkdale Cave and other caverns containing fossil bones in his "Reliquæ Diluvianæ."

time, to tear in pieces the body of any dead animal, and carry it to their den to feed upon it apart.

While Kirkdale Cave was evidently the haunt of Hyenas chiefly, other caverns in Germany and France were tenanted in a similar manner by a

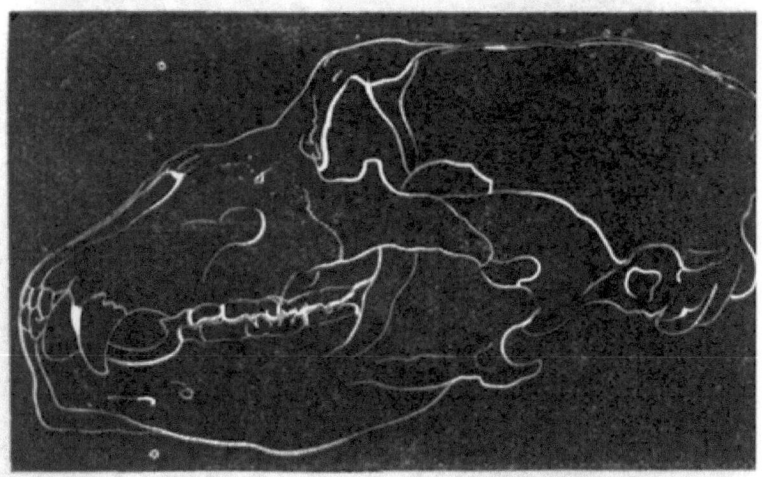

gigantic species of Bear. Their remains, mingled with those of the animals on which they fed, have been found in great numbers in the Cavern of Gailenreuth, in Franconia. The subjoined woodcut shows the head of this formidable beast, which must have exceeded in size any Bear now living. Indeed, although there were many smaller kinds, and the other types of the Animal Kingdom in the Tertiaries seem to approach very nearly both in size and general character their modern representatives, yet, on the whole, the earlier Mammalia were giants in comparison with those now living. The Mastodon and Mammoth, as compared with the modern Elephant, the Megatherium, as compared with the Sloths, or Ant-Eaters of present times, the Hyenas and Bears of the European caverns, and the fossil Elk of Ireland, by the side

of which even the Moose of our Northern woods is belittled, are remarkable instances in proof of this. One cannot but be struck with the fact that this first representation of Mammalia, the very impersonation of brute force in power, size, and ferocity, immediately preceded the introduction of man, with whose creation intelligence and moral strength became the dominant influences on earth.

Among those huge Tertiary Mammalia, one of those most common on the North-American continent seems to have been the Mastodon. The magnificent specimens preserved in this country are too well known to require description. The remains of the Rhinoceros occur also in the recent Tertiary deposits of North America, though as yet no perfect skeletons have been found. The Edentata, now confined to South America and the western coast of Africa, were also numerous in the Southern States during that time; their remains have been found as far north as the Salt Lick in Kentucky. But we must not judge of the Tertiary Edentata by any now known to us. The Sloths, the Armadillos, the Ant-Eaters, the Pangolins, are all animals of rather small size; but formerly they were represented by the gigantic Megatherium, the Megalonyx, and the Mylodon, some of which were larger than the Elephant, and others about the size of the Rhinoceros or Hippopotamus. The subjoined wood-cut represents a Mylodon in the act

Mylodon.

of lifting himself against the trunk of a tree. They were clumsy brutes, and though their limbs were evidently built with reference to powerful movements, perhaps climbing, or at least rising on their hind quarters, the act of climbing with them cannot have had anything of the nimbleness or activity generally associated with it. On the contrary, they probably were barely able to support their huge bodies on the hind limbs, which are exceedingly massive, and on the stiff, heavy tail, while they dragged down with their front limbs the branches of the trees, and fed upon them at leisure. The Zoölogical Museum at Cambridge is indebted to the generosity of Mr. Joshua Bates for a very fine set of casts taken from the Megatherium bones preserved in the British Museum and the College of Surgeons. They are now mounted, and may be seen in one of the exhibition-rooms of the building. Large Reptiles, but very unlike those of the Cretaceous and Jurassic epochs, belonging chiefly to the types of Turtles, Crocodiles, Pythons, and Salamanders, existed during the Tertiary epochs. The subjoined wood-cut represents a gigantic Salamander of the Tertiary deposits. It is a curious fact, illustrative of the ignorance of all anatomical science in those days, that, when the remains of this reptile (Andrias, as it is now called) were first discovered towards the close of the seventeenth century, they were described by old Professor Scheuchzer as the bones of an infant destroyed

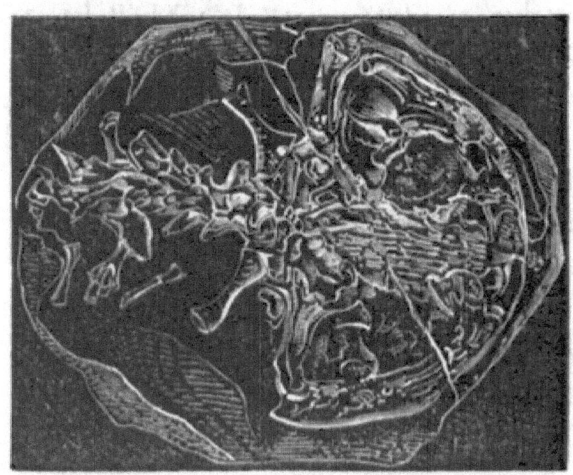

Andrias.

by the Deluge, and were actually preserved, not for their scientific value, but as precious relics of the Flood, and described in a separate pamphlet, entitled, "Homo Diluvii Testis." Among the Tertiary Reptiles the Turtles seem to have been a very prominent type, by their size as well as by their extensive distribution. Their remains have been found both in the far West and in the East. The fossil Turtles of Nebraska are well known to American naturalists; but the Oriental one exceeds them in size, and is, indeed, the most gigantic representative of the order known thus far. A man could stand under the arch of the shield of the old Himalayan Turtle preserved in the British Museum.

It would carry me too far, were I to attempt to give anything more than the most cursory sketch of the animals of the Tertiary age; and,

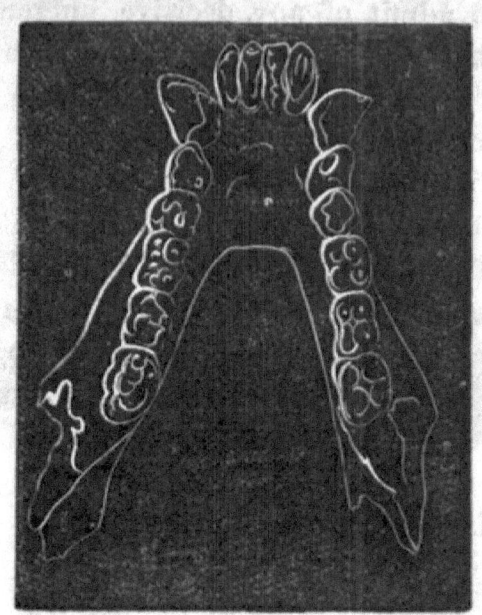

indeed, they are so well known, and have been so fully represented in text-books, that I fear some of my readers may think even now that I have dwelt too long upon them. Monkeys were unquestionably introduced upon earth before the close of the Tertiaries; some bones have been found in Southern France, and also on Mount Pentelicus in Greece, in the later Tertiary deposits; but these remains have not yet been collected in sufficient number to establish much more than the fact of their presence in the animal creation at that time. I do not offer any opinion respecting the fossil human bones so much discussed recently, because the evidence is at present too scanty to admit of any decisive judgment concerning them. It becomes, however, daily more probable that facts will force us sooner or later to admit that the creation of man lies far beyond any period yet assigned to it, and that a succession of human races, as of animals, have followed one another upon the earth. It may be the inestimable privilege of our young naturalists to solve this great problem, but the older men of our generation must be content to renounce this hope; we may have some prophetic vision of its fulfilment, we may look from afar into the land of promise, but we shall not enter in and possess it.

The other great types of the Animal Kingdom are very fully represented in the Tertiaries, and in their general appearance they approach much more closely those of the present creation than of any previous epochs. Professor Heer has collected and described the Tertiary Insects in great number and variety; and the Butterflies, Bugs, Flies, Grasshoppers, Dragon-Flies, Beetles, etc., described in his volumes, would hardly be distinguished from our own, except by a practised entomologist. Among Crustacea, the Shrimp-like forms of the earlier geological epochs have become much less conspicuous, while Crabs and Lobsters are now the prominent representatives of the class. Among Mollusks, the Chambered Shells, hitherto so numerous, have become, as they now are, very few in comparison with the naked Cephalopods. The Nautili, however, resemble those now living in the Pacific Ocean; and some fragments of the Paper-Nautilus have been found, showing that this delicate shell was already in existence. There is one very peculiar type of this class, belonging to the Tertiaries, which should not be passed by unnoticed. It partakes of the character both of the Cretaceous Belemnites and of the living Cuttle-Fish, and is known as the Spirulirostra. Another very char-

Spirulirostra.

acteristic group among the Tertiary Shells is that of the Nummulites, formerly placed by naturalists in immediate proximity with the Ammonites, on account of their internal partitions. This is now admitted to have been an error; their position is not yet fully determined, but they certainly stand very low in the scale, and have no affinity whatever with the Cephalopods. The subjoined wood-cut represents one of these Shells, so numerous in the Tertiaries that large masses of rock consist of their remains. The Univalve Shells or Gasteropods of the Tertiaries embraced all the families now living, including land

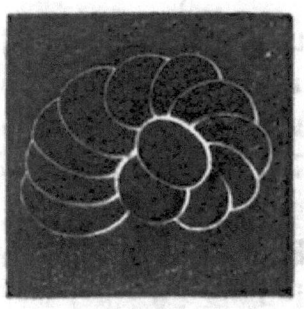
Nummulite.

and fresh-water Shells as well as the marine representatives of the type. Some of the latter, as, for instance, the Cerithium, are accumulated in vast numbers. The limestone quarries out of which Paris is chiefly built consist almost wholly of these Shells. The fresh-water basins were filled with Helices, one of which is represented in the following

Cerithium.

Helix.

wood-cut, with Planorbis, Limnæus, and other Shells resembling those now so common in all our lakes and rivers, and differing from the living ones only by slight specific characters.

The Bivalves also have the same resemblance to the present ones, including fresh-water Mussels, as the marine Clams and Oysters. Among Radiates, the higher Echini (Sea-Urchins) have become numerous, while the other Echinoderms of all families abound. Corals include, for the first time, the more highly organized Madrepores.

In the Tertiaries we see the dawn of the present condition of things, not only in the character of the animals and plants, but in the height of the mountains and in the distribution of land and sea.

Let us give a glance at the continents whose growth we have been following, and see what these more recent geological epochs have done for their completion. In Europe they have filled the basin in Central France, and converted all that region into dry land; they have filled also the channel between France and Spain; they have united Central Russia with the rest of Europe by the completion of Poland, and have greatly enlarged Austria and Turkey; they have completed the promontories of Italy and Greece, and have converted the inland sea at the foot of the Jura into the plain of Switzerland. But this fruitful period in the progress of the world, when the character of organic life was higher and the physical features of the earth more varied than ever before, was not without its storms and convulsions. The Pyrenees, the Apennines, the Alps, and with them the whole range of the Caucasus and Himalayas, were raised either immediately after the Cretaceous epoch, or in the course of the Tertiaries. Indeed, with this most significant passage in her history, Europe acquired all her essential characters. There remained, it is true, much to be done in what is called by geologists "modern times." The work of the artist is not yet finished when his statue is blocked out and the grand outline of his conception stands complete; and there still remained, after the earth was rescued from the water, after her framework of mountains was erected, after her soil was clothed with field and forest, processes by which her valleys were to be made more fruitful, her gulfs to be filled with the rich detritus poured into them by the rivers, her whole surface to be rendered more habitable for the higher races who were to possess it.

We left America at the close of the Carboniferous epoch. A glance at the geological map will show the reader that during the Permian, Triassic, and Jurassic epochs little was added to the United States, though here and there deposits belonging to each of them crop out. In the Cretaceous epoch, however, large tracts of land were accumulated, chiefly in the South and West; and during the Tertiaries the continent was very nearly completed, leaving only a narrow gulf running up to the neighborhood of St. Louis to be filled by modern detritus, and the peninsula of Florida to be built by the industrious Coral-Workers of our own period. The age of the Alleghany chain is not yet positively determined, but it was probably raised at the close of the Carboniferous epoch. Up to that time, only the Laurentian Hills, the northern side of that mountainous triangle which now makes the skeleton, as it were, of the United States, existed. The upheaval of the Alleghanies added its eastern side, raising the central part of the continent so as to form a long slope from the base of the Alleghanies to the Pacific Ocean; but it was not until the Tertiary Age that the upheaval of the great chain at the West completed the triangle, and transformed that wide westerly slope into the Mississippi Valley, bounded on one side by the Alleghanies, and on the other by the Rocky Mountains.

It is my belief, founded upon the tropical character of the Fauna, that a much milder climate then prevailed over the whole northern hemisphere than is now known to it. Some naturalists have supposed that the presence of the tropical Mammalia in the Northern Temperate Zone might be otherwise accounted for, — that they might have been endowed with warmer covering, with thicker hair or fur. But I think the simpler and more natural reason for their existence throughout the North is to be found in the difference of climate; and I am the more inclined to this opinion because the Tertiary animals generally, the Fishes, Shells, etc., in the same regions, are more closely allied in character to those now living in the Tropics than to those of the Temperate Zones. The Tertiary age may be called the geological summer; we shall see, hereafter, how abruptly it was brought to a close.

One word more as to the relation of the Tertiary Mammalia to the creation which preceded them. I can only repeat here the argument used before: the huge quadrupeds characteristic of these epochs make their appearance suddenly, and the deposits containing them follow as immediately upon those of the Cretaceous epoch, in which no trace of them occurs, as do those of the Cretaceous upon those of the Jurassic epoch. I would remind the reader that in the central basin of France, in which Cuvier found his first Palæotherium, and which afterwards proved to have been thickly settled by the early Mammalia, the deposits of the Jurassic, Cretaceous, and Tertiary epochs follow each other in immediate, direct, uninterrupted succession; that the same is true of other localities, in Germany, in Southern Europe, in England, where the most complete collections have been made from all these deposits; and there has never been brought to light a single fact leading us to suppose that any intermediate forms have ever existed through which more recent types have been developed out of older ones. For thirty years Geology has been gradually establishing, by evidence the fulness and accuracy of which are truly amazing, the regularity in the sequence of the geological formations, and distinguishing, with ever-increasing precision, the specific differences of the animals and plants contained in these accumulations of past ages. These results bear living testimony to the wonderful progress of the kindred sciences of Geology and Palæontology in the last half-century; and the development-theory has but an insecure foundation so long as it attempts to strengthen itself by belittling the geological record, the assumed imperfection of which, in default of positive facts, has now become the favorite argument of its upholders.

VIII.

THE FORMATION OF GLACIERS.

The long summer was over. For ages a tropical climate had prevailed over a great part of the earth, and animals whose home is now beneath the Equator roamed over the world from the far South to the very borders of the Arctics. The gigantic quadrupeds, the Mastodons, Elephants, Tigers, Lions, Hyenas, Bears, whose remains are found in Europe from its southern promontories to the northernmost limits of Siberia and Scandinavia, and in America from the Southern States to Greenland and the Melville Islands, may indeed be said to have possessed the earth in those days. But their reign was over. A sudden intense winter, that was also to last for ages, fell upon our globe; it spread over the very countries where these tropical animals had their homes, and so suddenly did it come upon them that they were embalmed beneath masses of snow and ice, without time even for the decay which follows death. The Elephant whose story was told at length in the preceding article was by no means a solitary specimen; upon further investigation it was found that the disinterment of these large tropical animals in Northern Russia and Asia was no unusual occurrence. Indeed, their frequent discoveries of this kind had given rise among the ignorant inhabitants to the singular superstition already alluded to, that gigantic moles lived under the earth, which crumbled away and turned to dust as soon as they came to the upper air. This tradition, no doubt, arose from the fact, that, when in digging they came upon the bodies of these animals, they often found them perfectly preserved under the frozen ground, but the moment they were exposed to heat and light they decayed and fell to pieces at once. Admiral Wrangel, whose Arctic explorations have been so valuable to science, tells us that the remains of these animals are heaped up in such quantities in certain parts of Siberia that he and his men climbed over ridges and mounds consisting entirely of the bones of Elephants, Rhinoceroses, etc. From these facts it would seem that they roamed over all these northern regions in troops as large and numerous as the Buffalo herds that wander over our Western prairies now. We are indebted to Russian naturalists, and especially to Rathke, for the most minute investigations of these remains, in which even the texture of the hair, the skin, and flesh has been subjected by him to microscopic examination as accurate as if made upon any living animal.

We have as yet no clew to the source of this great and sudden change of climate. Various suggestions have been made, — among others, that formerly the inclination of the earth's axis was greater, or that a submersion of the continents under water might have produced a decided increase of cold; but none of these explanations are satisfactory, and science has yet to find any cause which accounts for all the phenomena connected with it. It seems, however, unquestionable, that since the opening of the Tertiary age a cosmic summer and winter have succeeded each other, during which a Tropical heat and an Arctic cold have alternately prevailed over a great portion of the present Temperate Zone. In the so-called drift (a superficial deposit subsequent to the Tertiaries, of the origin of which I shall speak presently) there are found far to the south of their present abode the remains of animals whose home now is in the Arctics or the coldest parts of the Temperate Zones. Among them are the Musk-Ox, the Reindeer, the Walrus, the Seal, and many kinds of Shells characteristic of the Arctic regions. The northernmost part of Norway and Sweden is at this day the southern limit of the Reindeer in Europe; but their fossil remains are found in large quantities in the drift about the neighborhood of Paris, and quite recently they have been traced even to the foot of the Pyrenees, where their presence would, of course, indicate a climate similar to the one now prevailing in Northern Scandinavia. Side by side with the remains of the Reindeer are found those of the European Marmot, whose present home is in the mountains, about six thousand feet above the level of the sea. The occurrence of these animals in the superficial deposits of the plains of Central Europe, one of which is now confined to the high North, and the other to mountain-heights, certainly indicates an entire change of climatic conditions since the time of their existence. European Shells now confined to the Northern Ocean are found as fossils in Italy,— showing that, while the present Arctic climate prevailed in the Temperate Zone, that of the Temperate Zone extended much farther south to the regions we now call sub-tropical. In America there is abundant evidence of the same kind; throughout the recent marine deposits of the Temperate Zone, covering the low lands above tide-water on this continent, are found fossil Shells whose present home is on the shores of Greenland. It is not only in the Northern hemisphere that these remains occur, but in Africa and in South America, wherever there has been an opportunity for investigation, the drift is found to contain the traces of animals whose presence indicates a climate many degrees colder than that now prevailing there.

But these organic remains are not the only evidence of the geological winter. There are a number of phenomena indicating that during this period two vast caps of ice stretched from the Northern pole southward and from the Southern pole northward, extending in each case far toward the Equator, — and that ice-fields, such as now spread over the Arctics, covered a great part of the Temperate Zones, while the line of perpetual ice and snow in the tropical mountain-ranges descended far below its present limits. As the explanation of these facts has been drawn from the study of glacial action, I shall devote this and subsequent articles to some account of glaciers and of the phenomena connected with them.

The first essential condition for the formation of glaciers in mountain-ranges is the shape of their valleys. Glaciers are by no means in proportion to the height and extent of mountains. There are many mountain-chains as high or higher than the Alps, which can boast of but few and small glaciers, if, indeed, they have any. In the Andes, the Rocky Mountains, the Pyrenees, the Caucasus, the few glaciers remaining from the great ice-period are insignificant in size. The volcanic, cone-like shape of the Andes gives indeed, but little chance for the formation of glaciers, though their summits are capped with snow. The glaciers of the Rocky Mountains have been little explored, but it is known that they are by no means extensive. In the Pyrenees there is but one great glacier, though the height of these mountains is such, that, were the shape of their valleys favorable to the accumulation of snow, they might present beautiful glaciers. In the Tyrol, on the contrary, as well as in Norway and Sweden, we find glaciers almost as fine as those of Switzerland, in mountain-ranges much lower than either of the above-named chains. But they are of diversified forms, and have valleys widening upward on the slope of long crests. The glaciers on the Caucasus are very small in proportion to the height of the range; but on the northern side of the Himalaya there are large and beautiful ones, while the southern slope is almost destitute of them. Spitzbergen and Greenland are famous for their extensive glaciers, coming down to the sea-shore, where huge masses of ice, many hundred feet in thickness, break off and float away into the ocean as icebergs. At the Aletsch in Switzerland, where a little lake lies in a deep cup between the mountains, with the glacier coming down to its brink, we have these Arctic phenomena on a small scale; a miniature iceberg may often be seen to break off from the edge of the larger mass, and float out upon the surface of the water. Icebergs were first traced back to their true origin by the nature of the land-ice of which they are always composed, and which is quite distinct in structure and consistency from the marine ice produced by frozen sea-water, and called "ice-flow" by the Arctic explorers, as well as from the pond or river ice, resulting from the simple congelation of fresh water, the laminated structure of which is in striking contrast to the granular structure of glacier ice.

Water is changed to ice at a certain temperature under the same law of crystallization by which any inorganic bodies in a fluid state may assume a solid condition, taking the shape of perfectly regular crystals, which combine at certain angles with mathematical precision. The frost does not form a solid, continuous sheet of ice over an expanse of water, but produces crystals, little ice-blades, as it were, which shoot into each other at angles of thirty or sixty degrees, forming the closest net-work. Of course, under the process of alternate freezing and thawing, these crystals lose their regularity, and soon become merged in each other. But even then a mass of ice is not continuous or compact throughout, for it is rendered completely porous by air-bubbles, the presence of which is easily explained. Ice being in a measure transparent to heat, the water below any frozen surface is nearly as susceptible to the elevation of the temperature without as if it were in immediate contact with it. Such changes of temperature produce air-bubbles, which float upward against the lower surface of the ice and are stranded there. At night there may come a severe frost; new ice is then formed below the air-bubbles, and they are thus caught and imprisoned, a layer of air-bubbles between two layers of ice, and this process may be continued until we have a succession of such parallel layers, forming a body of ice more or less permeated with air. These air-bubbles have the power also of extending their own area, and thus rendering the whole mass still more porous; for, since the ice offers little or no obstacle to the passage of heat, such an air-bubble may easily become heated during the day; the moment it reaches a temperature above thirty-two degrees, it melts the ice around it, thus clearing a little space for itself, and rises through the water produced by the action of its own warmth. The spaces so formed are so many vertical tubes in the ice, filled with water, and having an air-bubble at the upper extremity.

Ice of this kind, resulting from the direct congelation of water, is easily recognized under all circumstances by its regular stratification, the alternate beds varying in thickness according to the intensity of the cold, and its continuance below the freezing-point during a longer or shorter period. Singly, these layers consist of irregular crystals confusedly blended together, as in large masses of crystalline rocks in which a crystalline structure prevails, though regular crystals occur but rarely. The appearance of stratification is the result of the circumstances under which the water congeals. The temperature varies much more rapidly in the atmosphere around the earth than in the waters upon its surface. When the atmosphere above any sheet of water sinks below the freezing-point, there stretches over its surface a stratum of cold air, determining by its intensity and duration the formation of the first stratum of ice. According to the alternations of temperature, this process goes on with varying activity until the sheet of ice is so thick that it becomes itself a shelter to the water below, and protects it, to a certain degree, from the cold without. Thus a given thickness of ice may cause a suspension of the freezing process, and the first ice-stratum may even be partially thawed before the cold is renewed with such intensity as to continue the thickening of the ice-sheet by the addition of fresh layers. The strata or beds of ice increase gradually in this manner, their separation being rendered still more distinct by the accumulation of air-bubbles, which, during a warm and clear day, may rise from a muddy bottom in great numbers. In consequence of these occasional collections of air-bubbles, the layers differ, not only in density and closeness, but also in color, the more compact strata being blue and transparent, while those containing a greater quantity of air-bubbles are opaque and whitish, like water beaten to froth.

A cake of pond-ice, such as is daily left in summer at our doors, if held against the light and turned in different directions, will exhibit all these phenomena very distinctly, and we may learn still more of its structure by watching its gradual melting. The process of decomposition is as different in fresh-water ice and in land- or glacier-ice as that of their formation. Pond-ice, in contact with warm air, melts uniformly over its whole surface, the mass being thus gradually reduced from the exterior till it vanishes completely. If the process be slow, the temperature of the air-bubbles contained in it may be so raised as to form the vertical funnels or tubes alluded to above. By the anastomosing of these funnels, the whole mass may be reduced to a collection of angular pyramids, more or less closely united by cross-beams of ice, and it finally falls to pieces when the spaces in the interior have become so numerous as to render it completely cavernous.

Such a breaking-up of the ice is always caused by the enlargement of the open spaces produced by the elevated temperature of the air-bubbles, these spaces being necessarily more or less parallel with one another, and vertical in their position, owing to the natural tendency of the air-bubbles to work their way upward till they reach the surface, where they escape. A sheet of ice, of this kind, floating upon water, dissolves in the same manner, melting wholly from the surface, if the process be sufficiently rapid, or falling to pieces, if the air-bubbles are gradually raised in their temperature sufficiently to render the whole mass cavernous and incoherent. If we now compare these facts with what is known of the structure of land-ice, we shall see that the mode of formation in the two cases differs essentially.

Land-ice, of which both the ice-fields of the Arctics and the glaciers consist, is produced by the slow and gradual transformation of snow into ice; and though the ice thus formed may eventually be as clear and transparent as the purest pond- or river-ice, its structure is nevertheless entirely distinct. We may compare these different processes during any moderately cold winter in the ponds and snow-meadows immediately about us. We need not join an Arctic exploring expedition, nor even undertake a more tempting trip to the Alps, in order to investigate these phenomena for ourselves, if we have any curiosity to do so. The first warm day after a thick fall of light, dry snow, such as occurs in the coldest of our winter weather, is sufficient to melt its surface. As this snow is porous, the water readily penetrates it, having also a tendency to sink by its own weight, so that the whole mass becomes more or less filled with moisture in the course of the day. During the lower temperature of the night, however, the water is frozen again, and the snow is now filled with new ice-particles. Let this process be continued long enough, and the mass of snow is changed to a kind of ice-gravel, or, if the grains adhere together, to something like what we call pudding-stone, allowing, of course, for the difference of material; the snow, which has been rendered cohesive by the process of partial melting and regelation, holding the ice-globules together, just as the loose materials of the pudding-stone are held together by the cement which unites them.

Within this mass, air is intercepted and held inclosed between the particles of ice. The process by which snow-flakes or snow-crystals are transformed into grains of ice, more or less compact, is easily understood. It is the result of a partial thawing, under a temperature maintained very nearly at thirty-two degrees, falling sometimes a little below, and then rising a little above the freezing-point, and thus producing constant alternations of freezing and thawing in the same mass of snow. This process amounts to a kind of kneading of the snow, and when combined with the cohesion among the particles more closely held together in one snow-flake, it produces granular ice. Of course, the change takes place gradually, and is unequal in its progress at different depths in the same bed of recently fallen snow. It depends greatly on the amount of moisture infiltrating the mass, whether derived from the melting of its own surface, or from the accumulation of dew or the falling of rain or mist upon it. The amount of water retained within the mass will also be greatly affected by the bottom on which it rests and by the state of the atmosphere. Under a certain temperature, the snow may only be glazed at the surface by the formation of a thin, icy crust, an outer membrane, as it were, protecting the mass below from a deeper transformation into ice; or it may be rapidly soaked throughout its whole bulk, the snow being thus changed into a kind of soft pulp, what we commonly call slosh, which, upon freezing, becomes at once compact ice; or, the water sinking rapidly, the lower layers only may be soaked, while the upper portion remains comparatively dry. But, under all these various circumstances, frost will transform the crystalline snow into more or less compact ice, the mass of which will be composed of an infinite number of aggregated snow-particles, very unequal in regularity of outline, and cemented by ice of another kind, derived from the freezing of the infiltrated moisture, the whole being interspersed with air. Let the temperature rise, and such a mass, rigid before, will resolve itself again into disconnected ice-particles, like grains more or less rounded. The process may be repeated till the whole mass is transformed into very compact, almost uniformly transparent and blue ice, broken only by the intervening air-bubbles. Such a mass of ice, when exposed to a temperature sufficiently high to dissolve it, does not melt from the surface and disappear by a gradual diminution of its bulk, like pond-ice, but crumbles into its original granular fragments, each one of which melts separately. This accounts for the sudden disappearance of icebergs, which, instead of slowly dissolving into the ocean, are often seen to fall to pieces and vanish at once.

Ice of this kind may be seen forming every winter on our sidewalks, on the edge of the little ditches which drain them, or on the summits of broad gate-posts when capped with snow. Of such ice glaciers are composed; but, in the glacier, another element comes in which we have not considered as yet, — that of immense pressure in consequence of the vast accumulations of snow within circumscribed spaces. We see the same effects produced on a small scale, when snow is transformed into a snowball between the hands. Every boy who balls a mass of snow in his hands illustrates one side of glacial phenomena. Loose snow, light and porous, and pure white from the amount of air contained in it, is in this way presently converted into hard, compact, almost transparent ice. This change will take place sooner, if the snow be damp at first, — but if dry, the action of the hand will presently produce moisture enough to complete the process. In this case, mere pressure produces the same effect which, in the cases we have been considering above, was brought about by alternate thawing and freezing, — only, that in the latter the ice is distinctly granular, instead of being uniform throughout, as when formed under pressure. In the glaciers we have the two processes combined. But the investigators of glacial phenomena have considered too exclusively one or the other: some of them attributing glacial motion wholly to the dilatation produced by the freezing of infiltrated moisture in the mass of snow; others accounting for it entirely by weight and pressure. There is yet a third class, who, disregarding the real properties of ice, would have us believe, that, because tar, for instance, is viscid when it moves, therefore ice is viscid because it moves. We shall see hereafter that the phenomena exhibited in the onward movement of glaciers are far more diversified than has generally been supposed.

There is no chain of mountains in which the shape of the valleys is more favorable to the formation of glaciers than the Alps. Contracted at their lower extremity, these valleys widen upward, spreading into deep, broad, trough-like depressions. Take, for instance, the valley of Hassli, which is not more than half a mile wide where you enter it above Meyringen; it opens gradually upward, till, above the Grimsel, at the foot of the Finster-Aarhorn, it measures several miles across. These huge mountain-troughs form admirable cradles for the snow, which collects in immense quantities within them, and, as it moves slowly down from the upper ranges, is transformed into ice on its way, and compactly crowded into the narrower space below. At the lower extremity of the glacier the ice is pure, blue, and transparent, but, as we ascend, it appears less compact, more porous and granular, assuming gradually the character of snow, till in the higher regions the snow is as light, as shifting, and incoherent, as the sand of the desert. A snow-storm on a mountain-summit is very different from a snow-storm on the plain, on account of the different degrees of moisture in the atmosphere. At great heights, there is never dampness enough to allow the fine snow-crystals to coalesce and form what are called "snow-flakes." I have even stood on the summit of the Jungfrau when a frozen cloud filled the air with ice-needles, while I could see the same cloud pouring down sheets of rain upon Lauterbrunnen below. I remember this spectacle as one of the most impressive I have witnessed in my long experience of Alpine scenery. The air immediately about me seemed filled with rainbow-dust, for the ice-needles glittered with a thousand hues under the decomposition of light upon them, while the dark storm in the valley below offered a strange contrast to the brilliancy of the upper region in which I stood. One wonders where even so much vapor as may be transformed into the finest snow should come from at such heights. But the warm winds, creeping up the sides of the valleys, the walls of which become heated during the middle of the day, come laden with moisture which is changed to a dry snow like dust as soon as it comes into contact with the intense cold above.

Currents of warm air affect the extent of the glaciers, and influence also the line of perpetual snow, which is by no means at the same level, even in neighboring localities. The size of glaciers, of course, determines to a great degree the height at which they terminate, simply because a small mass of ice will melt more rapidly, and at a lower temperature, than a larger one. Thus, the small glaciers, such as those of the Rothhorn or of Trift, above the Grimsel, terminate at a considerable height above the plain, while the Mer de Glace, fed from the great snow-caldrons of Mont Blanc, forces its way down to the bottom of the valley of Chamouni, and the glacier of Grindelwald, constantly renewed from the deep reservoirs where the Jungfrau hoards her vast supplies of snow, descends to about four thousand feet above the sea-level. But the glacier of the Aar, though also very large, comes to a pause at about six thousand feet above the level of the sea; for the south wind from the other side of the Alps, the warm sirocco of Italy, blows across it, and it consequently melts at a higher level than either the Mer de Glace or the Grindelwald. It is a curious fact, that in the valley of Hassli the temperature frequently rises instead of falling as you ascend; at the Grimsel, the temperature is at times higher than at Meyringen below, where the warmer winds are not felt so directly. The glacier of Aletsch, on the southern slope of the Jungfrau, and into which many other glaciers enter, terminates also at a considerable height, because it turns into the valley of the Rhone, through which the southern winds blow constantly.

Under ordinary conditions, vegetation fades in these mountains at the height of six thousand feet, but, in consequence of prevailing winds, and the sheltering influence of the mountain-walls, there is no uniformity in the limit of perpetual snow and ice. Where currents of warm air are very constant, glaciers do not occur at all, even where other circumstances are favorable to their formation. There are valleys in the Alps far above six thousand feet which have no glaciers, and where perpetual snow is seen only on their northern sides. These contrasts in temperature lead to the most wonderful contrasts in the aspect of the soil; summer and winter lie side by side, and bright flowers look out from the edge of snows that never melt. Where the warm winds prevail, there may be sheltered spots at a height of ten or eleven thousand feet, isolated nooks opening southward where the most exquisite flowers bloom in the midst of perpetual snow and ice; and occasionally I have seen a bright little flower with a cap of snow over it that seemed to be its shelter. The flowers give, indeed, a peculiar charm to these high Alpine regions. Occurring often in beds of the same kind, forming green, blue or yellow patches, they seem nestled close together in sheltered spots, or even in fissures and chasms of the rock, where they gather in dense quantities. Even in the sternest scenery of the Alps some sign of vegetation lingers; and I remember to have found a tuft of lichen growing on the only rock which pierced through the ice on the summit of the Jungfrau. It was a species then unknown to botanists, since described under the name of Umbelicarus Higinis. The absolute solitude, the intense stillness of the upper Alps is most impressive; no cattle, no pasturage, no bird, nor any sound of life, — and, indeed, even if there were, the rarity of the air in these high regions is such that sound is hardly transmissible. The deep repose, the purity of aspect of every object, the snow, broken only by ridges of angular rocks, produce an effect no less beautiful than solemn. Sometimes, in the midst of the wide expanse, one comes upon a patch of the so-called red snow of the Alps. At a distance, one would say that such a spot marked some terrible scene of blood, but, as you come nearer, the hues are so tender and delicate, as they fade from deep red to rose, and so die into the pure colorless snow around, that the first impression is completely dispelled. This red snow is an organic growth, a plant springing up in such abundance that it colors extensive surfaces, just as the microscopic plants dye our pools with green in the spring. It is an *Alga (Protocoites nivalis*) well known in the Arctics, where it forms wide fields in the summer.

With the above facts before us concerning the materials of which glaciers are composed, we may now proceed to consider their structure more fully in connection with their movements and the effects they produce on the surface over which they extend. It has already been stated that the ice of the glaciers has not the same appearance everywhere, but differs according to the level at which it stands. In consequence of this we distinguish three very distinct regions in these frozen fields, the uppermost of which, upon the sides of the steepest and highest slopes of the mountain-ridges, consists chiefly of layers of snow piled one above another by the successive snow-falls of the colder seasons, and which would remain in uniform superposition but for the change to which they are subjected in consequence of a gradual downward movement, causing the mass to descend by slow degrees, while new accumulations in the higher regions annually replace the snow which has been thus removed to an inferior level. We shall consider hereafter the process by which this change of position is brought about. For the present it is sufficient to state that such a transfer, by which a balance is preserved in the distribution of the snow, takes place in all glaciers, so that, instead of increasing indefinitely in the upper regions, where on account of the extreme cold there is little melting, they permanently preserve about the same thickness, being yearly reduced by their downward motion in a proportion equal to their annual increase by fresh additions of snow. Indeed, these reservoirs of snow maintain themselves at the same level, much as a stream, into which many rivulets empty, remains within its usual limits in consequence of the drainage of the average supply. Of course, very heavy rains or sudden thaws at certain seasons or in particular years may cause an occasional overflow of such a stream; and irregularities of the same kind are observed during certain years or at different periods of the same year in the accumulations of snow, in consequence of which the successive strata may vary in thickness. But in ordinary times layers from six to eight feet deep are regularly added annually to the accumulation of snow in the higher regions, — not taking into account, of course, the heavy drifts heaped up in particular localities, but estimating the uniform average increase over wide fields. This snow is gradually transformed into more or less compact ice, passing through an intermediate condition analogous to the slosh of our roads, and in that condition chiefly occupies the upper part of the extensive troughs into which these masses descend from the loftier heights. This region is called the region of the *névé*. It is properly the birthplace of the glaciers, for it is here that the transformation of the snow into ice begins. The *névé* ice, though varying in the degree of its compactness and solidity, is always very porous and whitish in color, resembling somewhat frozen slosh, while lower down in the region of the glacier proper the ice is close, solid, transparent, and of a bluish tint.

But besides the difference in solidity and in external appearance, there are also many other important changes taking place in the ice of these different regions, to which we shall return presently. Such modifications arise chiefly from the pressure to which it is subjected in its downward progress, and to the alterations, in consequence of this displacement, in the relative position of the snow- and ice-beds, as well as to the influence exerted by the form of the valleys themselves, not only upon the external aspect of the glaciers, but upon their internal structure also. The surface of a glacier varies greatly in character in these different regions. The uniform even surfaces of the upper snow-fields gradually pass into a more undulating outline, the pure white fields become strewn with dust and sand in the lower levels, while broken bits of stone and larger fragments of rock collect upon them, which assume a regular arrangement, and produce a variety of features most startling and incomprehensible at first sight, but more easily understood when studied in connection with the whole series of glacial phenomena. They are then seen to be the consequence of the general movement of the glacier, and of certain effects which the course of the seasons, the action of the sun, the rain, the reflected heat from the sides of the valley, or the disintegration of its rocky walls, may produce upon the surface of the ice. In the next article we shall consider in detail all these phenomena, and trace them in their natural connection. Once familiar with these facts, it will not be difficult correctly to appreciate the movement of the glacier and the cause of its inequalities. We shall see, that, in consequence of the greater or less rapidity in the movement of certain portions of the mass, its centre progressing faster than its sides, and the upper, middle, and lower regions of the same glacier advancing at different rates, the strata which in the higher ranges of the snow-fields were evenly spread over wide expanses, become bent and folded to such a degree that the primitive stratification is nearly obliterated, while the internal mass of the ice has also assumed new features under these new circumstances. There is, indeed, as much difference between the newly formed beds of snow in the upper region and the condition of the ice at the lower end of a glacier as between a recent deposit of coral sand or a mud-bed in an estuary and the metamorphic limestone or clay slate twisted and broken as they are seen in the very chains of mountains from which the glaciers descend. A geologist, familiar with all the changes to which a bed of rock may be subjected from the time it was deposited in horizontal layers up to the time when it was raised by Plutonic agencies along the sides of a mountain-ridge, bent and distorted in a thousand directions, broken through the thickness of its mass, and traversed by innumerable fissures which are themselves filled with new materials, will best be able to understand how the stratification of snow may be modified by pressure and displacement so as finally to appear like a laminated mass full of cracks and crevices, in which the original stratification is recognized only by the practical student. I trust in my next article I shall be able to explain intelligibly to my readers even these extreme alterations in the condition of the primitive snow of the Alpine summits.

IX.

INTERNAL STRUCTURE AND PROGRESSION OF GLACIERS.

It is not my intention, in these articles, to discuss a general theory of the glaciers upon physical and mechanical principles. My special studies, always limited to Natural History, have but indifferently fitted me for such a task, and quite recently the subject has been admirably treated from this point of view by Dr. Tyndall, in his charming volume entitled "Glaciers of the Alps." I have worked upon the glaciers as an amateur, devoting my summer vacations, with friends desirous of sharing my leisure, to excursions in the Alps, for the sake of relaxation from the closer application of my professional studies, and have considered them especially in their connection with geological phenomena, with a view of obtaining, by means of a thorough acquaintance with glaciers as they exist now, some insight into the glacial phenomena of past times, the distribution of drift, the transportation of boulders, etc. It was, however, impossible to treat one series of facts without some reference to the other; but such explanations as I have given of the mechanism of the glacier, in connection with its structure, are presented in the language of the unprofessional observer, without any attempt at the technicalities of the physicist. I do not wonder, therefore, that those who have looked upon the glacier chiefly with reference to the physical and mechanical principles involved in its structure and movement should have found my Natural Philosophy defective. I am satisfied with their agreement as to my correct observation of the facts, and am the less inclined to quarrel with the doubts thrown on my theory since I see that the most eminent physicists of the day do not differ from me more sharply than they do from each other. The facts will eventually test all our theories, and they form, after all, the only impartial jury to which we can appeal. In the mean while, I am not sorry that just at this moment, when recent investigations and publications have aroused new interest in the glaciers, the course of these articles brings me naturally to a discussion of the subject in its bearing upon geological questions. I shall, however, address myself especially, as I have done throughout these papers, to my unprofessional readers, who, while they admire the glaciers, may also wish to form a general idea of their structure and mode of action, as well as to know something of the important part they have played in the later geological history of our earth. It would, indeed, be out of place, were I to undertake here a discussion of the different views entertained by the various students who have investigated the glacier itself, among whom Dr. Tyndall is especially distinguished, or those of the more theoretical writers, among whom Mr. Hopkins occupies a prominent position.

Removed, as I am, from all possibility of renewing my own observations, begun in 1836 and ended in 1845, I will take this opportunity to call the attention of those particularly interested in the matter to one essential point with reference to which all other observers differ from me. I mean the stratification of the glacier, which I do not believe to be rightly understood, even at this moment. It may seem presumptuous to dissent absolutely from the statements of one who has seen so much and so well as Dr. Tyndall, on a question for the solution of which, from the physicist's point of view, his special studies have been a far better preparation than mine; and yet I feel confident that I was correct in describing the stratification of the glacier as a fundamental feature of its structure, and the so-called dirt-bands as the margins of the snow-strata successively deposited, and in no way originating in the ice-cascades. I shall endeavor to make this plain to my readers in the course of the present article. I believe, also, that renewed observations will satisfy dissenting observers that there really exists a net-work of capillary fissures extending throughout the whole glacier, constantly closing and reopening, and constituting the channels by means of which water filtrates into its mass. This infiltration, also, has been denied, in consequence of the failure of some experiments in which an attempt was made to introduce colored fluids into the glacier. To this I can only answer, that I succeeded completely, myself, in the self-same experiments which a later investigator found impracticable, and that I see no reason why the failure of the latter attempt should cast a doubt upon the former. The explanation of the difference in the result may, perhaps, be found in the fact, that, as a sponge gorged with water can admit no more fluid than it already contains, so the glacier, under certain circumstances, and especially at noonday in summer, may be so soaked with water that all attempts to pour colored fluids into it would necessarily fail. I have stated, in my work upon glaciers, that my infiltration experiments were chiefly made at night; and I chose that time, because I knew the glacier would most readily admit an additional supply of liquid from without when the water formed during the day at its surface and rushing over it in myriad rills had ceased to flow.

While we admit a number of causes as affecting the motion of a glacier,— namely, the natural tendency of heavy bodies to slide down a sloping surface, the pressure to which the mass is subjected forcing it onward, the infiltration of moisture, its freezing and consequent expansion, — we must also remember that these various causes, by which the accumulated masses of snow and ice are brought down from higher to lower levels, are not all acting at all times with the same intensity, nor is their action always the same at every point of the moving mass. While the bulk of snow and ice moves from higher to lower levels, the whole mass of the snow, in consequence of its own downward tendency, is also under a strong vertical pressure, arising from its own incumbent weight, and that pressure is, of course, greater at its bottom than at its centre or surface. It is therefore plain, that, inasmuch as the snow can be compressed by its own weight, it will be more compact at the bottom of such an accumulation than at its surface, this cause acting most powerfully at the upper part of a glacier, where the snow has not yet been transformed into a more solid icy mass. To these two agencies, the downward tendency and the vertical pressure, must be added the pressure from behind, which is most effective where the mass is largest and the amount of motion in a given time greatest. In the glacier, the mass is, of course, largest in the centre, where the trough which holds it is deepest, and least on the margins, where the trough slopes upward and becomes more shallow. Consequently, the middle of a glacier always advances more rapidly than the sides.

Were the slope of the ground over which it passes, combined with the pressure to which the mass is subjected, the whole secret of the onward progress of a glacier, it is evident that the rate of advance would be gradually accelerated, reaching its maximum at its lower extremity, and losing its impetus by degrees on the higher levels, nearer the point where the descent begins. This, however, is not the case. The glacier of the Aar, for instance, is about ten miles in length; its rate of annual motion is greatest near the point of junction of the two great branches by which it is formed, diminishing farther down, and reaching a minimum at its lower extremity. But in the upper regions, near their origin, the progress of these branches is again gradually less.

Let us see whether the next cause of displacement, the infiltration of moisture, may not in some measure explain this retardation, at least of the lower part of the glacier. This agency, like that of the compression of the snow by its own weight and the pressure from behind, is most effective where the accumulation is largest. In the centre, where the body of the mass is greatest, it will imbibe the most moisture. But here a modifying influence comes in, not sufficiently considered by the investigators of glacial structure. We have already seen that snow and ice, at different degrees of compactness, are not equally permeable to moisture. Above the line at which the annual winter snow melts, there is, of course, little moisture; but below that point, as soon as the temperature rises in summer sufficiently to melt the surface, the water easily penetrates the mass, passing through it more readily where the snow is lightest and least compact, — in short, where it has not begun its transformation into ice. A summer's day sends countless rills of water trickling through such a mass of snow. If the snow be loose and porous throughout, the water will pass through its whole thickness, accumulating at the bottom, so that the lower portion of the mass will be damper, more completely soaked with water, than the upper part; if, on the contrary, in consequence of the process previously described, alternate melting and freezing combined with pressure, the mass has assumed the character of icy snow, it does not admit moisture so readily, and still farther down, where the snow is actually transformed into pure compact ice, the amount of surface-water admitted into its structure will, of course, be greatly diminished. There may, however, be conditions under which even the looser snow is comparatively impervious to water; as, for instance, when rain falls upon a snow-field which has been long under a low temperature, and an ice-crust is formed upon its surface, preventing the water from penetrating below. Admitting, as I believe we must, that the water thus introduced into the snow and ice is one of the most powerful agents to which its motion is due, we must suppose that it has a twofold influence, since its action when fluid and when frozen would be different. When fluid, it would contribute to the advance of the mass in proportion to its quantity; but when frozen, its expansion would produce a displacement corresponding to the greater volume of ice as compared with water; add to this that while trickling through the mass it will loosen and displace the particles of already consolidated ice. I have already said that I did not intend to trespass on the ground of the physicist, and I will not enter here upon any discussion as to the probable action of the laws of hydrostatic pressure and dilatation in this connection. I will only state, that, so far as my own observation goes, the movement of the glacier is most rapid where the greatest amount of moisture is introduced into the mass, and that I believe there must be a direct relation between these two facts. If I am right in this, then the motion, so far as it is connected with infiltrated moisture or with the dilatation caused by the freezing of that moisture, will, of course, be most rapid where the glacier is most easily penetrated by water, namely, in the region of the *névé* and in the upper portion of the glacier-troughs, where the *névé* begins to be transformed into more or less porous ice. This cause also accounts, in part at least, for another singular fact in the motion of the glacier: that, in its higher levels, where its character is more porous and the water entering at the surface sinks readily to the bottom, there the bottom seems to move more rapidly than the superficial parts of the mass, whereas, at the lower end of the glacier, in the region of the compact ice, where the infiltration of the water at the bottom is at its minimum, while the disintegrating influences at the surface admit of infiltration to a certain limited depth, there the motion is greater near the surface than toward the bottom. But, under all circumstances, it is plain that the various causes producing motion, gravitation, pressure, infiltration of water, frost, will combine to propel the mass at a greater rate along its axis than near its margins. For details concerning the facts of the case, I would refer to my work entitled " Système Glaciaire."

We will next consider the stratification of the glacier. I have stated, in my introductory remarks, that I consider this to be one of its primary and fundamental features, and I confess, that, after a careful examination of the results obtained by my successors in the field of glacial phenomena, I still believe that the original stratification of the mass of snow from which the glacier arises gives us the key to many facts of its internal structure. The ultimate features resulting from this connection are so exceedingly intricate and entangled that their relation is not easily explained. Nevertheless, I trust my readers will follow me in this Alpine excursion, where I shall try to smooth the asperities of the road for them as much as possible.

Imparted to it, at the very beginning of its formation, by the manner in which snow accumulates, and retained through all its transformations, the stratification of a glacier, however distorted, and at times almost obliterated, remains, notwithstanding, as distinct to one who is acquainted with all its phases, as is the strat ified character of metamorphic rocks to the skilful geologist, even though they may be readily mistaken for plutonic masses by the common observer. Indeed, even those secondary features, as the dirt-bands, for instance, which we shall see to be intimately connected with snow-strata, and which eventually become so prominent as to be mistaken for the cause of the lines of stratification, do nevertheless tend, when properly understood, to make the evidence of stratification more permanent, and to point out its primitive lines.

On the plains, in our latitude, we rarely have the accumulated layers of several successive snow-storms preserved one above another. We can, therefore, hardly imagine with what distinctness the sequence of such beds is marked in the upper Alpine regions. The first cause of this distinction between the layers is the quality of the snow when it falls, then the immediate changes it undergoes after its deposit, then the falling of mist or rain upon it, and lastly and most efficient of all, the accumulation of dust upon its surface. One who has not felt the violence of a storm in the high mountains, and seen the clouds of dust and sand carried along with the gusts of wind passing over a mountain-ridge and sweeping through the valley beyond, can hardly conceive that not only the superficial aspect of a glacier, but its internal structure also, can be materially affected by such a cause. Not only are dust and sand thus transported in large quantities to the higher mountain regions, but leaves are frequently found strewn upon the upper glacier, and even pine-cones, and maple-seeds flying upward on their spread wings, are scattered thousands of feet above and many miles beyond the forests where they grow.

This accumulation of sand and dust goes on all the year round, but the amount accumulated over one and the same surface is greatest during the summer, when the largest expanse of rocky wall is bare of snow, and its loose soil dried by the heat so as to be easily dislodged. This summer deposit of loose inorganic materials, light enough to be transported by the wind, forms the main line of division between the snow of one year and the next, though only that of the last year is visible for its whole extent. Those of the preceding years, as we shall see hereafter, exhibit only their edges cropping out lower down one beyond another, being brought successively to lower levels by the onward motion of the glacier.

Other observers of the glacier, Professor Forbes and Dr. Tyndall, have noticed only the edges of these seams, and called them dirt-bands. Looking upon them as merely superficial phenomena, they have given explanations of their appearance which I hold to be quite untenable. Indeed, to consider these successive lines of dirt on the glacier as limited only to its surface, and to explain them from that point of view, is much as if a geologist were to consider the lines presented by the strata on a cut through a sedimentary mass of rock as representing their whole extent, and to explain them as a superficial deposit due to external causes.

A few more details may help to make this statement clearer to my readers. Let us imagine that a fresh layer of snow has fallen in these mountain regions, and that a deposit of dirt has been scattered over its surface, which, if any moisture arises from the melting of the snow or from the falling of rain or mist, will become more closely compacted with it. The next snow-storm deposits a fresh bed of snow, separated from the one below it by the sheet of dust just described, and this bed may, in its turn, receive a like deposit. For greater ease and simplicity of explanation, I speak here as if each successive snow-layer were thus indicated; of course this is not literally true, because snow-storms in the winter may follow each other so fast that there is no time for such a collection of foreign materials upon each newly formed surface. But whenever such a fresh snow-bed, or accumulation of beds, remains with its surface exposed for some time, such a deposit of dirt will inevitably be found upon it. This process may go on till we have a number of successive snow-layers divided from each other by thin sheets of dust. Of course, such seams, marking the stratification of snow, are as permanent and indelible as the seams of coarser materials alternating with the finest mud in a sedimentary rock.

The gradual progress of a glacier, which, though more rapid in summer than in winter, is never intermitted, changes the relation of these beds to each other. Their lower edge is annually cut off at a certain level, because the snow deposited every winter melts with the coming summer, up to a certain line, determined by the local climate of the place. But although the snow does not melt above this line, we have seen, in the preceding article, that it is prevented from accumulating indefinitely in the higher regions by its own tendency to move down to the lower valleys, and crowding itself between their walls, thus to force its way toward the outlet below. Now, as this movement is very gradual, it is evident that there must be a perceptible difference in the progress of the successive layers, the lower and older ones getting the advance of the upper and more recent ones: that is, when the snow that has covered the face of the country during one winter melts away from the glacier up to the so-called snow-line, there will be seen cropping out below and beyond that line the layers of the preceding years, which are already partially transformed into ice, and have become a part of the frozen mass of the glacier with which they are moving onward and downward. In the autumn, when the dust of a whole season has been accumulated upon the service of the preceding winter's snow, the extent of the layer which year after year will henceforth crop out lower down, as a dirt-band, may best be appreciated.

Beside the snow-layers and the sheets of dust alternating with them, there is still another feature of the horizontal and parallel structure of the mass in immediate connection with those above considered. I allude to the layers of pure compact ice occurring at different intervals between the snow-layers. In July, when the snow of the preceding winter melts up to the line of perpetual snow, the masses above, which are to withstand the summer heat and become part of the glacier forever, or at least until they melt away at the lower end, begin to undergo the changes through which all snow passes before it acquires the character of glacial ice. It thaws at the surface, is rained upon, or condenses moisture, thus becoming gradually soaked, and after assuming the granular character of *névé*-ice, it ends in being transformed into pure compact ice. Toward the end of August, or early in September, when the nights are already very cold in the Alps, but prior to the first permanent autumnal snow-falls, the surface of these masses becomes frozen to a greater or less depth, varying, of course, according to temperature. These layers of ice become numerous and are parallel to each other, like the layers of ice formed from slosh. Such crusts of ice I have myself observed again and again upon the glacier. This stratified snowy ice is now the bottom on which the first autumnal snow-falls accumulate. These sheets of ice may be formed not only annually before the winter snows set in, but may recur at intervals whenever water accumulating upon an extensive snow-surface, either in consequence of melting or of rain, is frozen under a sharp frost before another deposit of snow takes place. Or suppose a fresh layer of light porous snow to have accumulated above one, the surface of which has already been slightly glazed with frost; rain or dew, falling upon the upper one, will easily penetrate it; but when it reaches the lower one, it will be stopped by the film of ice already formed, and, under a sufficiently low temperature, it will be frozen between the two. This result may be frequently noticed in winter, on the plains, where sudden changes of temperature take place.

There is still a third cause, to which the same result may possibly be due, and to which I shall refer at greater length hereafter; but as it has not, like the preceding ones, been the subject of direct observation, it must be considered as hypothetical. The admirable experiments of Dr. Tyndall have shown that water may be generated in ice by pressure, and it is therefore possible that at a lower depth in the glacier, where the incumbent weight of the mass above is suffiient to produce water, the water thus accumulated may be frozen into ice-layers. But this depends so much upon the internal temperature of the glacier, about which we know little beyond a comparatively superficial depth, that it cannot at present afford a sound basis even for conjecture.

There are, then, in the upper snow-fields three kinds of horizontal deposits: the beds of snow, the sheets of dust, and the layers of ice, alternating with each other. If, now, there were no modifying circumstances to change the outline and surface of the glacier, — if it moved on uninterruptedly through an open valley, the lower layers, forming the mass, getting by degrees the advance of the upper ones, our problem would be simple enough. We should then have a longitudinal mass of snow, enclosed between rocky walls, its surface crossed by straight transverse

11*

lines marking the annual additions to the glacier, as in the adjoining figure.

But that mass of snow, before it reaches the outlet of the valley, is to be compressed, contorted, folded, rent in a thousand directions. The beds of snow, which in the upper ranges of the mountain were spread out over broad, open surfaces, are to be crowded into comparatively circumscribed valleys, to force and press themselves through narrow passes, alternately melting and freezing, till they pass from the condition of snow into that of ice, to undergo, in short, constant transformations, by which the primitive stratifications will be extensively modified. In the first place, the more rapid motion of the centre of the glacier, as compared with the margins, will draw the lines of stratification downward toward the middle faster than at the sides. Accurate measurements have shown that the axis of a glacier may move ten- or twenty-fold more rapidly than its margins. This is not the place to introduce a detailed account of the experiments made to ascertain this result; but I would refer those who are interested in the matter to the measurements given in my "Système Glaciaire," where it will be seen that the middle may move at a rate of two hundred feet a year, while the margins may not advance more than fifteen or ten feet, or even less, in the same time. These observations of mine have the advantage over those of other observers, that, while they embrace the whole extent of the glacier, transversely as well as in its length, they cover a period of several successive years, instead of being limited to summer campaigns and a few winter observations. The consequence of this mode of progressing will be that the straight lines drawn transversely across the surface of the glacier above will be gradually changed to curved ones below. After a few years, such a line will appear on the surface of the glacier like a crescent, with the bow turned downward, within which, above, are other crescents, less and less sharply arched up to the last year's line, which may be again straight across the snow-field. (See the subjoined figure, which represents a part of the glacier of the Lauter-Aar.)

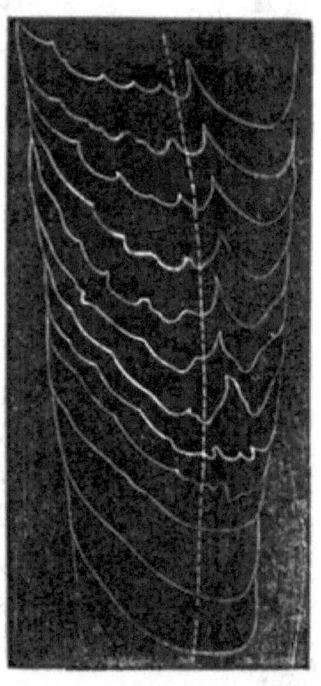

Thus the glacier records upon its surface its

annual growth and progress, and registers also the inequality in the rate of advance between the axis and the sides.

But these are only surface phenomena. Let us see what will be the effect upon the internal structure. We must not forget, in considering the changes taking place within glaciers, the shape of the valleys which contain them. A glacier lies in a deep trough, and the tendency of the mass will be to sink towards its deeper part, and to fold inward and downward, if subjected to a strong lateral pressure, — that is, to dip toward the centre and slope upward along the sides, following the scoop of the trough. If, now, we examine the face of a transverse cut in the glacier, we find it traversed by a number of lines, vertical in some places, more or less oblique in others, and frequently these lines are joined together at the lower ends, forming loops, some of which are close and vertical, while others are quite open. These lines are due to the folding of the strata in consequence of the lateral pressure they are subjected to when crowded into the lower course of the valleys, and the difference in their dip is due to the greater or less force of that pressure. The wood cut on the next page represents a transverse cut across the Lauter-Aar and the Finster-Aar, the two principal tributaries to the great Aar glacier, and includes also a number of small lateral glaciers which join them. The beds on the left, which dip least, and are only folded gently downward, forming very open loops, are those of the Lauter-Aar, where the lateral pressure is comparatively slight. Those which are almost vertical belong in part to the several small tributary glaciers, which have been crowded together and very strongly compressed, and partly to the Finster-Aar. The close uniform vertical lines in this wood-cut represent a different feature in the structure of the glacier, called blue bands, to which I shall refer presently. These loops or lines dipping into the internal mass of the glacier have been the subject of much discussion, and various theories have been recently proposed respecting them. I believe them to be caused, as I have said, by the snow-layers, originally deposited horizontally, but afterwards folded into a more or less vertical position, in consequence of the lateral pressure brought to bear upon them. The sheets of dust and of ice alternating with the snow-strata are

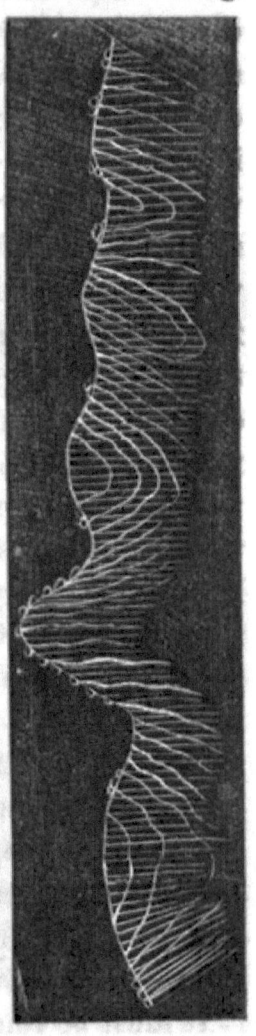

of course subjected to the same action, and are contorted, bent, and folded by the same lateral pressure.

Dr. Tyndall has advanced the view that the lines of apparent stratification, and especially the dirt-bands across the surface of the glacier, are due to ice-cascades: that is, the glacier, passing over a sharp angle, is cracked across transversely in consequence of the tension, and these rents, where the back of the glacier has been successively broken, when recompacted, cause the transverse lines, the dirt being collected in the furrow formed between the successive ridges. Unfortunately for his theory, the lines of stratification constantly occur in glaciers where no such ice-falls are found. His principal observations upon this subject were made on the Glacier du Géant, where the ice-cascade is very remarkable. The lines may perhaps be rendered more distinct on the Glacier du Géant by the cascade, and necessarily must be so, if the rents coincide with the limit at which the annual snow-line is nearly straight across the glacier. In the region of the Aar glacier, however, where my own investigations were made, all the tributaries entering into the larger glaciers are ribbed across in this way, and most of them join the main trunk over uniform slopes, without the slightest cascade.

It must be remembered that these surface-phenomena of the glacier are not to be seen at all times, nor under all conditions. During the first year of my sojourn on the glacier of the Aar, I was not aware that the stratification of its tributaries was so universal as I afterward found it to be; the primitive lines of the strata are often so far erased that they are not perceptible, except under the most favorable circumstances. But when the glacier has been washed clean by rain, and the light strikes upon it in the right direction, these lines become perfectly distinct, where, under different conditions, they could not be discerned at all. After passing many summers on the same glacier, renewing my observations year after year over the same localities, I can confidently state that not only do the lines of stratification exist throughout the great glacier of the Aar, but in all its tributaries also. Of course, they are greatly modified in the lower part of the glacier by the intimate fusion of its tributaries, and by the circumstance that their movement, primarily independent, is merged in the movement of the main glacier embracing them all. We have seen that not only does the centre of a glacier move more rapidly than its sides, but that the deeper mass of the glacier also moves at a different rate from its more superficial portion. My own observations (for the details of which I would again refer the reader to my "Système

Glaciaire") show that in the higher part of the glacier, especially in the region of the *névé*, the bottom of the mass seems to move more rapidly than the surface, while lower down, toward the

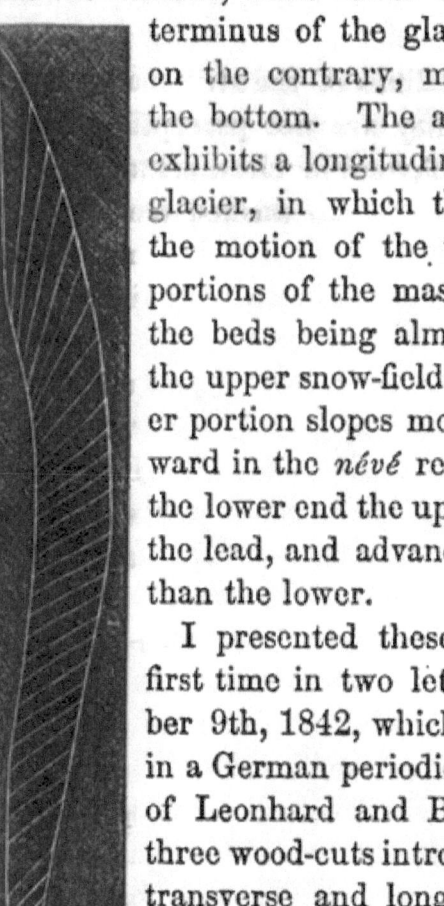

terminus of the glacier, the surface, on the contrary, moves faster than the bottom. The annexed wood-cut exhibits a longitudinal section of the glacier, in which this difference in the motion of the upper and lower portions of the mass is represented, the beds being almost horizontal in the upper snow-fields, while their lower portion slopes more rapidly downward in the *névé* region, and toward the lower end the upper portion takes the lead, and advances more rapidly than the lower.

I presented these results for the first time in two letters, dated October 9th, 1842, which were published in a German periodical, the Jahrbuch of Leonhard and Bronn. The last three wood-cuts introduced above, the transverse and longitudinal sections of the glacier, as well as that representing the concentric lines of stratification on the surface, are the identical ones contained in those communications.

These papers seem to have been overlooked by contemporary investigators, and I may be permitted to translate here a passage from one of them, since it sums up the results of the inequality of motion throughout the glacier and its influence on the primitive stratification of the mass in as few words and as correctly as I could give them to-day, twenty years later: — " Combining these views, it appears that the glacier may be represented as composed of concentric shells which arise from the parallel strata of the upper region by the following process. The primitively regular strata advance into gradually narrower and deeper valleys, in consequence of which the margins are raised, while the middle is bent not only downward, but, from its more rapid motion, forward also, so that they assume a trough-like form in the interior of the mass. Lower down, the glacier is worn by the surrounding air, and assumes the peculiar form characteristic of its lower course." The last clause alludes to another series of facts, which we shall examine in a future article, when we shall see that the heat of the walls in the lower part of its course melts the sides of the glacier, so that, instead of following the trough-like shape of the valley, it becomes convex, arching upward in the centre and sinking at the margins.

I have dwelt thus long, and perhaps my readers may think tediously, upon this part of my subject, because the stratification of the glacier has been constantly questioned by the more recent investigators of glacial phenomena, and has indeed been set aside as an exploded theory. They consider the lines of stratification, the dirt-bands, and the seams of ice alternating with the more porous snow, as disconnected surface-phenomena, while I believe them all to be intimately connected together as primary essential features of the original mass.

There is another feature of glacial structure, intimately connected, by similarity of position and aspect, with the stratification, which has greatly perplexed the students of glacial phenomena. I allude to the so-called blue bands, or bands of infiltration, also designated as veined structure, ribboned or laminated structure, marginal structure, and longitudinal structure. The difficulty lies, I believe, in the fact that two very distinct structures, that of the stratification and the blue bands, are frequently blended together in certain parts of the glacier in such a manner as to seem identical, while elsewhere the one is prominent and the other subordinate, and *vice versâ*. According to their various opportunities of investigation, observers have either confounded the two, believing them to be the same, or some have overlooked the one and insisted upon the other as the prevailing feature, while that very feature has been absolutely denied again by others who have seen its fellow only, and taken that to be the prominent and important fact in this peculiar structural character of the ice.

We have already seen how the stratification of the glacier arises, accompanied by layers of dust and other material foreign to the glacier, and how blue bands of compact ice may be formed parallel to the surface of these strata. We have also seen how the horizontality of these strata may be modified by pressure till they assume a position within the mass of the glacier, varying from a slightly oblique inclination to a vertical one. Now, while the position of the strata becomes thus altered under pressure, other changes take place in the constitution of the ice itself.

Before attempting to explain how these changes take place, let us consider the facts themselves. The mass of the glacier ice is traversed by thin bands of compact blue ice, these bands being very numerous along the margins of the glacier, where they constitute what Dr. Tyndall calls marginal structure, and still more crowded along the line upon which two glaciers unite, where he has called it longitudinal structure. In the latter case, where the extreme pressure resulting from the junction of two glaciers has rendered the strata nearly vertical, these blue bands follow their trend so closely that it is almost impossible to distinguish one from the other. It will be seen, on referring to the wood-cut on page 253, where the close, uniform, vertical lines represent the true veined structure, that at several points of that section the lines of stratification run so nearly parallel with them, that, were the former not drawn more strongly, they could not be easily distinguished from the latter. Along the margins, also, in consequence of the retarded motion, the blue bands and the lines of stratification run nearly parallel with each other, both following the sides of the trough in which they move.

Undoubtedly, in both these instances, we have two kinds of blue bands, namely: those formed primitively in a horizontal position, indicating seams of stratification, and those which have arisen subsequently in connection with the movement of the whole mass, which I have occasionally called bands of infiltration, as they appeared to me to be formed by the infiltration and freezing of water. The fact that these blue bands are most numerous where two glaciers are crowded together into a common bed naturally suggests pressure as their cause. And since the beautiful experiments of Dr. Tyndall have illustrated the internal liquefaction of ice by pressure, it becomes highly probable that his theory of the origin of these secondary blue bands is the true one. He suggests that layers of water may be formed in the glacier at right angles with the pressure, and pass into a state of solid ice upon the removal of that pressure, the pressure being of course relieved in proportion to the diminution in the body of the ice by compression. The number of blue bands diminishes as we recede from the source of the pressure, — few only being formed, usually at right angles with the surfaces of stratification, in the middle of a glacier, halfway between its sides. If they are caused by pressure, this diminution of their number toward the middle of the glacier would be inevitable, since the intensity of the pressure naturally fades as we recede from the motive power.

Dr. Tyndall also alludes to another structure of the same kind, which he calls transverse structure, where the blue bands extend in crescent-shaped curves, more or less arched, across the surface of the glacier. Where these do not coincide with the stratification, they are probably formed by vertical pressure in connection with the unequal movement of the mass.

With these facts before us, it seems to me plain that the primitive blue bands arise with the stratification of the snow in the very first formation of the glacier, while the secondary blue bands are formed subsequently, in consequence of the onward progress of the glacier and the pressure to which it is subjected. The secondary blue bands intersect the planes of stratification at every possible angle, and may therefore seem identical with the stratification in some places, while in others they cut it at right angles. It has been objected to my theory of glacial structure, that I have considered the so-called blue bands as a superficial feature when compared with the stratification. And in a certain sense this is true; since, if my views are correct, the glacier exists and is in full life and activity before the secondary blue bands arise in it, whereas the stratification is a feature of its embryo condition, already established in the accumulated snow before it begins its transformation into glacier-ice. In other words, the veined structure of the glacier is not a primary structural feature of its whole mass, but the result of various local influences acting upon the constitution of the ice; the marginal structure resulting from the resistance of the sides of the valley to the onward movement of the glacier, the longitudinal structure arising from the pressure caused by two glaciers uniting in one common bed, the transverse structure being produced by vertical pressure, in consequence of the weight of the mass itself and the increased rate of motion at the centre.

In the *névé* fields, where the strata are still

horizontal, the few blue bands observed are perpendicular to the strata of snow, and therefore also perpendicular to the blue seams of ice and the sheets of dust alternating with them. Upon the sides of the glacier they are more or less parallel to the slopes of the valley; along the line of junction of two glaciers they follow the vertical trend of the axis of the mass; while at intermediate positions they are more or less oblique. Along the outcropping edges of the strata, on the surface of the glacier, they follow more or less the dip of the strata themselves; that is to say, they are more or less parallel with the dirt-bands. In conclusion, I would recommend future investigators to examine the glaciers, with reference to the distribution of the blue bands, after heavy rains and during foggy days, when the surface is freed from the loose materials and decomposed fragments of ice resulting from the prolonged action of the sun.

The most important facts, then, to be considered with reference to the motion of the glacier are as follows. First, that the rate of advance between the axis and the margins of a glacier differs in the ratio of about ten to one and even less; that is to say, when the centre is advancing at a rate of two hundred and fifty feet a year, the motion toward the sides may be gradually diminished to two hundred, one hundred and fifty, one hundred, fifty feet, and so on, till nearest the margin it becomes almost inappreciable. Secondly, the rate of motion is not the same throughout the length of the glacier, the advance being greatest about half-way down in the region of the *névé*, and diminishing in rapidity both above and below; thus the onward motion in the higher portion of a glacier may not exceed twenty to fifty feet a year, while it reaches its maximum of some two hundred and fifty feet annually in the *névé* region, and is retarded again toward the lower extremity, where it is reduced to about one fourth of its maximum rate. Thirdly, the glacier moves at different rates throughout the thickness of its mass; toward the lower extremity of the glacier the bottom is retarded, and the surface portion moves faster, while in the upper region the bottom seems to advance more rapidly. I say *seems*, because upon this latter point there are no positive measurements, and it is only inferred from general appearances, while the former statement has been demonstrated by accurate experiments. Remembering the form of the troughs in which the glaciers arise, that they have their source in expansive, open fields of snow and *névé*, and that these immense accumulations move gradually down into ever-narrowing channels, though at times widening again to contract anew, their surface wasting so little from external influences that they advance far below the line of perpetual snow without any sensible diminution in size, it is evident that an enormous pressure must have been brought to bear upon them before they could have been packed into the lower valleys through which they descend.

Physicists seem now to agree that pressure is the chief agency in the motion of glaciers. No doubt, all the facts point that way; but it now becomes a matter of philosophical interest to determine in what direction it acts most powerfully, and upon this point glacialists are by no means agreed. The latest conclusion seems to be, that the weight of the advancing mass is itself the efficient cause of the motion. But while this is probably true in the main, other elements tending to the same result, and generally overlooked by investigators, ought to be taken into consideration; and before leaving the subject, I would add a few words upon infiltration in this connection.

The weight of the glacier, as a whole, is about the same all the year round. If, therefore, pressure, resulting from that weight, be the all-controlling agency, its progress should be uniform during the whole year, or even greatest in winter, which is by no means the case. By a series of experiments, I have ascertained that the onward movement, whatever be its annual average, is accelerated in spring and early summer. The average annual advance of the glacier being, at a given point, at the rate of about two hundred feet, its average summer advance, at the same point, will be at a rate of two hundred and fifty feet, while its average rate of movement in winter will be about one hundred and fifty feet. This can be accounted for only by the increased pressure due to the large accession of water trickling in spring and early summer into the interior through the network of capillary fissures pervading the whole mass. The unusually large infiltration of water at that season is owing to the melting of the winter snow. Careful experiments made on the glacier of the Aar, respecting the water thus accumulating on the surface, penetrating its mass, and finally discharged in part at its lower extremity, fully confirm this view. Here, then, is a powerful cause of pressure and consequent motion, quite distinct from the permanent weight of the mass itself, since it operates only at certain seasons of the year. In midwinter, when the infiltration is reduced to a minimum, the motion is least. The water thus introduced into the glacier acts, as we have seen above, in various ways: by its weight, by loosening the particles of snow and ice through which it trickles, and by freezing and consequent expansion, at least within the limits and during the season at which the temperature of the glacier sinks below 32° Fahrenheit. The simple fact, that in the spring the glacier swells on an average to about five feet more than its usual level, shows how important this infiltration must be. I can therefore only wonder that other glacialists have given so little weight to this fact. It is admitted by all, that the waste of a glacier at its surface, in consequence of evaporation and melting, amounts to about nine or ten feet in a year. At this rate of diminution, a glacier, even one thousand feet in thickness, could not advance during a single century without being exhausted. The water supplied by infiltration no doubt repairs the loss to a great degree. Indeed, the lower part of the glacier must be chiefly maintained from this source, since the annual increase from the fresh accumulations of snow is felt only above the snow-line, below which the yearly snow melts away and disappears. In a complete theory of the glaciers, the effect of so great an accession of plastic material cannot be overlooked.

I now come to some points in the structure of the glacier, the consideration of which is likely to have a decided influence in settling the conflicting views respecting their motion. The experiments of Faraday concerning regelation, and the application of the facts made known by the great English physicist to the theory of the glaciers, as first presented by Dr. Tyndall in his admirable work, show that fragments of ice with moist surfaces are readily reunited under pressure into a solid mass. It follows from these experiments, that glacier-ice, at a temperature of 32° Fahrenheit, may change its form and preserve its continuity during its motion, in virtue of the pressure to which it is subjected. The statement is, that, when two pieces of ice with moistened surfaces are placed in contact, they become cemented together by the freezing of a film of water between them, while, when the ice is below 32° Fahrenheit, and therefore *dry*, no effect of the kind can be produced. The freezing was also found to take place under water; and the result was the same, even when the water into which the ice was plunged was as hot as the hand can bear.

The fact that ice becomes cemented under these circumstances is fully established, and my own experiments have confirmed it to the fullest extent. I question, however, the statement, that regelation takes place *by the freezing of a film of water between the fragments.* I never have been able to detect any indication of the presence of such a film, and am, therefore, inclined to consider this result as akin to what takes place when fragments of moist clay or marl are pressed together and thus reunited. When examining beds of clay and marl, or even of compact limestone, especially in large mountain masses, I have frequently observed that the rock presents a network of minute fissures pervading the whole, without producing a distinct solution of continuity, though generally determining the lines according to which it breaks under sudden shocks. The network of capillary fissures pervading the glacier may fairly be compared to these rents in hard rocks; with this difference, however, that in ice they are more permeable to water than in stone.

How this network of capillary fissures is formed has not been ascertained by direct observation. Following, however, the transformation of the snow and *névé* into compact ice, it is easily conceived that the porous mass of snow, as it falls in the upper regions of the Alps, and in the broad caldrons in which the glaciers properly originate, cannot pass into solid ice, by the process described in a former article, without retaining within itself larger or smaller quantities of air. This air is finally surrounded from all sides by the cementation of the granules of *névé*, through the freezing of the water that penetrates it. So enclosed, the bubbles of air are subject to the same compression as the ice itself, and become more flattened in proportion as the snow has been more fully transformed into compact ice. As long as the transformation of snow into ice is not complete, a rise of its temperature to 32° Fahrenheit, accompanied with thawing, reduces it at once again to the condition of loose grains of *névé;* but when more compact, it always presents the aspect of a mass composed of angular fragments, wedged and dovetailed together, and separated by capillary fissures, the flattened air-bubbles trending in the same direction in each fragment, but varying in their trend from one fragment to another. There is, moreover, this important point to notice, — that, the older the *névé*, the larger are its composing granules; and where *névé* passes into porous ice, small angular fragments are mixed with rounded *névé*-granules, the angular fragments appearing larger and more numerous, and the *névé*-granules fewer, in proportion as the *névé*-ice has undergone most completely its transformation into compact glacier-ice. These facts show conclusively that the dimensions and form of the *névé*-granules, the size and shape of the angular fragments, the porosity of the ice, the arrangement of its capillary fissures, and the distribution and compression of the air-bubbles it contains, are all connected features, mutually dependent. Whether the transformation of snow into ice be the result of pressure only, or, as I believe, quite as much the result of successive thawings and freezings, these structural features can equally be produced, and exhibit these relations to one another. It may be, moreover, that, when the glacier is at a temperature below 32°, its motion produces extensive fissuration throughout the mass.

Now that water pervades this network of fissures in the glacier to a depth not yet ascertained, my experiments upon the glacier of the Aar have abundantly proved; and that the fissures themselves exist at a depth of two hundred and fifty feet I also know, from actual observation. All this can, of course, take place, even if the internal temperature of the glacier never should fall below 32° Fahrenheit; and it has actually been assumed that the temperature within the glacier does not fall below this point, and that, therefore, no phenomena, dependent upon a greater degree of cold, can take place beyond a very superficial depth, to which the cold outside may be supposed to penetrate. I have, however, observed facts which seem to me irreconcilable with this assumption. In the first place, a thermometrograph indicating —2° Centigrade (about 28° Fahrenheit) at a depth of a little over two metres, that is, about six feet and a half, has been recovered from the interior of the glacier of the Aar, while all my attempts to thaw out other instruments placed in the ice at a greater depth utterly failed, owing to the circumstance, that, after being left for some time in the glacier, they were invariably frozen up in newly formed water-ice, entirely different in its structure from the surrounding glacier-ice. This freezing could not have taken place, did the mass of the glacier never fall below 32° Fahrenheit. And this is not the only evidence of hard frost in the interior of the glaciers. The innumerable large walls of water-ice, which may be seen intersecting their mass in every direction and to any depth thus far reached, show that water freezes in their interior. It cannot be objected, that this is merely the result of pressure; since the thin fluid seams, exhibited under pressure in the interesting experiments of Dr. Tyndall, and described in his work under the head of Crystallization and Internal Liquefaction, cannot be compared to the large, irregular masses of water-ice found in the interior of the glacier, to which I here allude.

In the absence of direct thermometric observations, from which the lowest internal temperature of the glacier could be determined with precision in all its parts, we are certainly justified in assuming that every particle of water-ice found in the glacier, the formation of which cannot be ascribed to the mere fact of pressure, is due to the influence of a temperature inferior to 32° Fahrenheit at the time of its consolidation. The fact that the temperature in winter has been proved by actual experiment to fall as low as 28°

Fahrenheit, that is, four degrees below the freezing-point, at a depth of six feet below a thick covering of snow, though not absolutely conclusive as to the temperature at a greater depth, is certainly very significant.

Under these circumstances, it is not out of place to consider through what channels the low temperature of the air surrounding the glacier may penetrate into the interior. The heavy cold air may of course sink from the surface into every large open space, such as the crevasses, large fissures, and *moulins* or mill-like holes to be described in a future article; it may also penetrate with the currents which ingulf themselves under the glacier, or it may enter through its terminal vault, or through the lateral openings between the walls of the valley and the ice. Indeed, if all the spaces in the mass of the glacier, not occupied by continuous ice, could be graphically represented, I believe it would be seen that cold air surrounds the glacier-ice itself in every direction, so that probably no masses of a greater thickness than that already known to be permeable to cold at the surface would escape this contact with the external temperature. If this be the case, it is evident that water may freeze in any part of the glacier.

To substantiate this position, which, if sustained, would prove that the dilatation of the mass of the glacier is an essential element of its motion, I may allude to several other well-known facts. The loose snow of the upper regions is gradually transformed into compact ice. The experiments of Dr. Tyndall prove that this may be the result of pressure; but in the region of the *névé* it is evidently owing to the transformation of the snow-flakes into ice by repeated melting and freezing, for it takes place in the uppermost layers of the snow, where pressure can have no such effect, as well as in its deeper beds. I take it for granted, also, that no one, familiar with the presence of the numerous ice-seams parallel to the layers of snow in these upper regions of the glacier, can doubt that they, as well as the *névé*, are the result of frost. But be this as it may, the difference between the porous ice of the upper region of the glacier and the compact blue ice of its lower track seems to me evidence direct that at times the whole mass must assume the rigidity imparted to it by a temperature inferior to the freezing-point. We know that at 32° Fahrenheit, regelation renders the mass continuous, and that it becomes brittle only at a temperature below this. In other words, the ice can break up into a mass of disconnected fragments, such as the capillary fissures and the infiltration-experiments described in my " Système Glaciaire" show to exist, only when it is below 32° Fahrenheit. If it be contended that ice at 32° does break, and that therefore the whole mass of the glacier may break at that temperature, setting aside the contradiction to the facts of regelation which such an assumption involves, I would refer to Dr. Tyndall's experiments concerning the vacuous spots in the ice.

Those who have read his startling investigations will remember that by sending a beam of sunlight through ice he brought to view the primitive crystalline forms to which it owes its solidity, and that he insisted that these star-shaped figures are always in the plane of crystallization. Without knowing what might be their origin, I had myself noticed these fig-

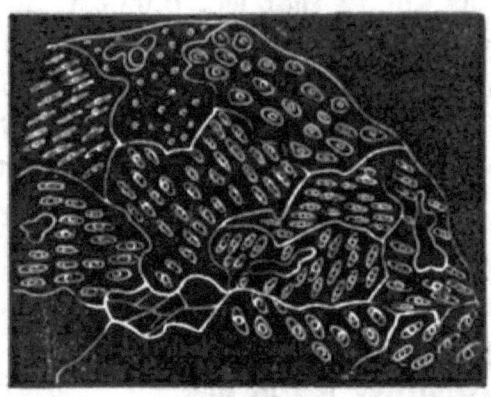

ures, and represented them in a diagram, part of which is reproduced in the accompanying wood-cut. I had considered them to be compressed air-bubbles; and though I cannot, under my present circumstances, repeat the experiment of Dr. Tyndall upon glacier-ice, I conceive that the star-shaped figures represented upon Pl. VII. Figs. 8 and 9, in my "Système Glaciaire," may refer to the same phenomenon as that observed by him in pond-ice. Yet while I make this concession, I still maintain, that besides these crystalline figures there exist compressed air-bubbles in the angular fragments of the glacier-ice, as shown in the preceding wood-cut; and that these bubbles are grouped in sets, trending in the same direction in one and the same fragment, and diverging under various angles in the different fragments. I have explained this fact concerning the position of the compressed air-bubbles, by assuming that ice, under various pressure, may take the appearance it presents in each fragment with every compressed air-bubble trending in the same direction, while their divergence in the different fragments is owing to a change in the respective position of the fragments resulting from the movement of the whole glacier. I have further assumed, that throughout the glacier the change of the snow and porous ice into compact ice is the result of successive freezing, alternating with melting, or at least with the resumption of a temperature of 32°

Fahrenheit in consequence of the infiltration of liquid water, to which the effects of pressure must be added, the importance of which in this connection no one could have anticipated prior to the experiments of Dr. Tyndall. Of course, if the interior temperature of the glacier never falls below 32°, the changes here alluded to could not take place. But if the *vacuous spaces* observed by Dr. Tyndall are really identical with the spaces I have described as *extremely flattened air-bubbles*, I think the arrangement of these spaces as above described proves that it freezes in the interior of the glacier to the depth at which these crosswise fragments have been observed; that is, at a depth of two hundred feet. For, since the experiments of Dr. Tyndall show that the vacuous spaces are parallel to the surface of crystallization, and as no crystallization of water can take place unless the surrounding temperature fall below 32°, it follows that these vacuous spaces could not exist in such large continuous fragments, presenting throughout the fragments the same trend, if there had been no frost within the mass, affecting the whole of such a fragment while it remained in the same position.

The most striking evidence, in my opinion, that at times the whole mass of the glacier actually freezes, is drawn from the fact, already alluded to, that, while the surface of the glacier loses annually from nine to ten feet of its thickness by evaporation and melting, it swells, on the other hand, in the spring, to the amount of about five feet. Such a dilatation can hardly be the result of pressure and the packing of the snow and ice, since the difference in the bulk of the ice brought down, during one year, from a point above to that under observation, would not account for the swelling. It is more readily explained by the freezing of the water of infiltration during spring and early summer, when the infiltration is most copious and the winter cold has been accumulating for the longest time. This view of the case is sustained by Élie de Beaumont, who states his opinion upon this point as follows:—

" Pendant l'hiver, la température de la surface du glacier s'abaisse à un grand nombre de degrés au-dessous de zéro, et cette basse température pénètre, quoique avec un affaiblissement graduel, dans l'intérieur de la masse. Le glacier se fendille par l'effet de la contraction résultant de ce refroidissement. Les fentes restent d'abord vides, et concourent au refroidissement des glaciers en favorisant l'introduction de l'air froid extérieur; mais au printemps, lorsque les rayons du soleil échauffent la surface de la neige qui couvre le glacier, ils la ramènent d'abord à zéro, et ils produisent ensuite de l'eau à zéro qui tombe dans le glacier refroidi et fendillé. Cette eau s'y congèle à l'instant, en laissant dégager de la chaleur qui tend à ramener le glacier à zéro ; et le phénomène se continue jusqu'à ce que la masse entière du glacier refroidi soit ramené à la température de zéro."*

But where direct observations are still so scanty, and the interpretations of the facts so conflicting, it is the part of wisdom to be circumspect in forming opinions. This much, however, I believe to be already settled: that any theory which ascribes the very complicated phenomena of the glacier to one cause must be defective and one-sided. It seems to me most

---

* "During the winter, the temperature at the surface of the glacier sinks a great many degrees below 32° Fahrenheit, and this low temperature penetrates, though at a gradually decreasing rate, into the interior of the mass. The glacier becomes fissured in consequence of the contraction resulting from this cooling process. The cracks remain open at first, and contribute to lower the temperature of the glacier by favoring the introduction of the cold air from without; but in the spring, when the rays of the sun raise the temperature of the snow covering the glacier, they first bring it back to 32° Fahrenheit, and presently produce water at 32°, which falls into the chilled and fissured mass of the glacier. There this water is instantly frozen, releasing heat which tends to bring back the glacier to the temperature of 32°; and this process continues till the entire mass of the cooled glacier returns to the temperature of 32°."

probable, that, while pressure has the larger share in producing the onward movement of the glacier, as well as in the transformation of the snow into ice, a careful analysis of all the facts will show that this pressure is owing partly to the weight of the mass itself, partly to the pushing on of the accumulated snow from behind, partly to its sliding along the surface upon which it rests, partly to the weight of water pervading the whole, partly to the softening of the rigid ice by the infiltration of water, and partly, also, to the dilatation of the mass, resulting from the freezing of this water. These causes, of course, modify the ice itself, while they contribute to the motion. Further investigations are required to ascertain in what proportion these different influences contribute to the general result, and at what time and under what circumstances they modify most directly the motion of the glacier.

That a glacier cannot be altogether compared to a river, although there is an unmistakable analogy between the flow of the one and the onward movement of the other, seems to me plain, — since the river, by the combination of its tributaries, goes on increasing in bulk in consequence of the incompressibility of water, while a glacier gradually thins out in consequence of the packing of its mass, however large and numerous may be its accessions. The analogy fails also in one important point, that of the acceleration of speed with the steepness of the slope. The motion of the glacier bears no such direct relation to the inclination of its bed. And though in a glacier, as in a river, the axis of swiftest motion is thrown alternately on one or the other side of the valley, according to its shape and slope, the very nature of ice makes it impossible that eddies should be formed in the glacier, and the impressive feature of whirlpools is altogether wanting in them. What have been called glacier-cascades bear only a remote resemblance to river-cascades, as in the former the surface only is thrown into confusion by breaking, without affecting the primitive structure;* and I reiterate my formerly expressed opinion, that even the stratification of the upper regions is still recognizable at the lower end of the glacier of the Rhone.

The internal structure of the glacier has already led me beyond the limits I had proposed to myself in the present article. But I trust my readers will not be discouraged by this dry discussion of various theories concerning it, and will meet me again on the glacier, when we will examine together some of its more picturesque

* For the evidence of this statement I must, however, refer to my work on Glaciers, already so often quoted in this article, where it may be found with all the necessary details.

features, its crevasses, its rivulets and cascades, its moraines, its boulders, etc., and endeavor also to track its ancient course and boundaries in earlier geological times.

X.

EXTERNAL APPEARANCE OF GLACIERS.

Thus far we have examined chiefly the internal structure of the glacier; let us look now at its external appearance, and at the variety of curious phenomena connected with the deposit of foreign materials upon its surface, some of which seem quite inexplicable at first sight. Among the most striking of these are the large boulders elevated on columns of ice, standing sometimes ten feet or more above the level of the glacier, and the sand-pyramids, those conical hills of sand which occur not infrequently on the larger Alpine glaciers. One is at first quite at a loss to explain the presence of these pyramids in the midst of a frozen ice-field, and yet it has a very simple cause.

I have spoken of the many little rills arising on the surface of the ice in consequence of its melting. Indeed, the voice of the waters is rarely still on the glacier during the warm season, except at night. On a summer's day, a thousand streams are born before noontide, and die again at sunset; it is no uncommon thing to see a full cascade come rushing out from the lower end of a glacier during the heat of the day, and vanish again at its decline. Suppose one of these rivulets should fall into a deep, circular hole, such as often occur on the glacier, and the nature of which I shall presently explain, and that this cylindrical opening narrows to a mere crack at a greater or less depth within the ice, the water will find its way through the crack and filter down into the deeper mass; but the dust and sand carried along with it will be caught there, and form a deposit at the bottom of the hole. As day after day, throughout the summer, the rivulet is renewed, it carries with it an additional supply of these light materials, until the opening is gradually filled and the sand is brought to a level with the surface of the ice. We have already seen, that, in consequence of evaporation, melting, and other disintegrating causes, the level of the glacier sinks annually at the rate of from five to ten feet, according to stations. The natural consequence, of course, must be, that the sand is left standing above the surface of the ice, forming a mound which would constantly increase in height in proportion to the sinking of the surrounding ice, had it sufficient solidity to retain its original position. But a heap of sand, if unsupported, must very soon subside and be dispersed; and, indeed, these pyramids, which are often quite lofty, and yet look as if they would crumble at a touch, prove, on nearer examination, to be perfectly solid, and are, in fact, pyramids of ice with a thin sheet of sand spread over them. A word will explain how this transformation is brought about. As soon as the level of the glacier falls below the sand, thus depriving it of support, it sinks down and spreads slightly over the surrounding surface. In this condition it protects the ice immediately beneath it from the action of the sun. In proportion as the glacier wastes, this protected area rises above the general mass and becomes detached from it. The sand, of course, slides down over it, spreading toward its base, so as to cover a wider space below, and an ever-narrowing one above, until it gradually assumes the pyramidal form in which we find it, covered with a thin coating of sand. Every stage of this process may occasionally be seen upon the same glacier, in a number of sand-piles raised to various heights above the surface of the ice, approaching the perfect pyramidal form, or falling to pieces after standing for a short time erect.

The phenomenon of the large boulders, supported on tall pillars of ice, is of a similar character. A mass of rock, having fallen on the surface of the glacier, protects the ice immediately beneath it from the action of the sun; and as the level of the glacier sinks all around it, in consequence of the unceasing waste of the surface, the rock is gradually left standing on an ice-pillar of considerable height. In proportion as the column rises, however, the rays of the sun reach its sides, striking obliquely upon them under the boulder, and wearing them away, until the column becomes at last too slight to sustain its burden, and the rock falls again upon the glacier; or, owing to the unequal action of the sun, striking of course with most power on the southern side, the top of the pillar becomes slanting, and the boulder slides off. These ice-pillars, crowned with masses of rock, form a very picturesque feature in the scenery of the glacier, and are represented in many of the landscapes in which Swiss artists have endeavored to reproduce the grandeur and variety of Alpine views, especially in the masterly Aquarelles of Lory. The English reader will find them admirably well described and illustrated in Dr. Tyndall's work upon the glaciers. They are known throughout the Alps as "glacier-tables;" and many a time my fellow-travellers and I have spread our frugal meal on such a table, erected, as it seemed, especially for our convenience.

Another curious effect is that produced by small stones or pebbles, small enough to become heated through by the sun in summer. Such a heated pebble will of course melt the ice below it, and so wear a hole for itself into which it sinks. This process will continue as long as the sun reaches the pebble with force enough to heat it. Numbers of such deep, round holes, like organ-pipes, varying in size from the diameter of a minute pebble or a grain of coarse sand to that of an ordinary stone, are found on the glacier, and at the bottom of each is the pebble by which it was bored. The ice formed by the freezing of water collecting in such holes and in the fissures of the surface is a pure crystallized ice, very different in color from the ice of the great mass of the glacier produced by snow; and sometimes, after a rain and frost, the surface of a glacier looks like a mosaic-work, in consequence of such veins and cylinders or spots of clear ice with which it is inlaid.

Indeed, the aspect of the glacier changes constantly with the different conditions of the temperature. We may see it, when, during a long dry season, it has collected upon its surface all sorts of light floating materials, as dust, sand, and the like, so that it looks dull and soiled, — or when a heavy rain has washed the surface clean from all impurities and left it bright and fresh. We may see it when the heat and other disintegrating influences have acted upon the ice to a certain superficial depth, so that its surface is covered with a decomposed crust of broken, snowy ice, so permeated with air that it has a dead-white color, like pounded ice or glass. Those who see the glacier in this state miss the blue tint so often described as characteristic of its appearance in its lower portion, and as giving such a peculiar beauty to its caverns and vaults. But let them come again after a summer storm has swept away this loose sheet of broken, snowy ice above, and before the same process has had time to renew it, and they will find the compact, solid surface of the glacier of as pure a blue as if it reflected the sky above. We may see it in the early dawn, before the new ice of the preceding night begins to yield to the action of the sun, and the surface of the glacier is veined and inlaid with the water poured into its holes and fissures during the day, and transformed into pure, fresh ice during the night, — or when the noonday heat has wakened all its streams, and rivulets sometimes as large as rivers rush along its surface, find their way to the lower extremity of the glacier, or, dashing down some gaping crevasse or open well, are lost beneath the ice.

It would seem, from the quantity of water that is sometimes ingulfed within these open breaks in the ice, that the glacier must occasionally be fissured to a very great depth. I remember once, when boring a hole in the glacier in order to let down a self-regulating thermometer into its interior, seeing an immense fissure suddenly rent open, in consequence, no doubt, of the shocks given to the ice by the blows of the instruments. The effect was like that of an earthquake; the mass seemed to rock beneath us, and it was difficult to keep our feet. One of these glacial rivers was flowing past the spot at the time, and it was instantly lost in the newly formed chasm. However deep and wide the fissure might be, such a stream of water, constantly poured into it, and daily renewed throughout the summer, must eventually fill it and overflow, unless it finds its way through the whole mass of the glacier to the bottom on which it rests; — it must have an outlet above or below. The fact that considerable rivulets (too broad to leap across, and too deep to wade through safely even with high boots) may entirely vanish in the glacier unquestionably shows one of two things, — that the whole mass must be soaked with water like a wet sponge, or the cavities must reach the bottom of the glacier. Probably the two conditions are generally combined.

In direct connection with the narrower fissures are the so-called *moulins*, — the circular wells on the glacier already alluded to when speaking of the sand-hills. We will suppose that a trans verse, narrow fissure has been formed across the glacier, and that one of the many rivulets flowing longitudinally along its surface empties into it. As the surface-water of the glacier producing these rivulets arises not only from the melting of the ice but also from the condensation of vapor, or even from rain-falls, and flows over the scattered dust-particles and fragments of rock, it has always a temperature slightly above $32°$, so that such a rivulet is necessarily warmer than the icy edge of the fissure over which it precipitates itself. In consequence of its higher temperature it melts the edge, gradually wearing it backward, till the straight margin of the fissure at the spot over which the water falls is changed to a semicircle; and, as much of the water dashes in spray and foam against the other side, the same effect takes place there, by which a corresponding semicircle is formed exactly opposite the first. This goes on not only at the upper margin, but through the whole depth of the opening as far down as the water carries its higher temperature. In short, a semicircular groove is excavated on either side of the fissure for its whole depth along the line on which the rivulet holds its downward course. After a time, in consequence of the motion of the glacier, such a fissure may close again, and then the two semicircles thus brought together form at once one continuous circle, and we have one of the round deep openings on the glacier known as *moulins*, or wells, which may of course become perfectly dry if any accident turns the rivulet aside or dries up its source. The most common cause of the intermittence of such a waterfall is the formation of a crevasse higher up, across the water course which supplied it, and which now begins another excavation.

These wells are often very profound. I have lowered a line for more than seven hundred feet in one of them before striking bottom; and one is by no means sure even then of having sounded the whole depth, for it may often happen that the water meets with some obstacle which prevents its direct descent, and, turning aside, continues its deeper course at a different angle. Such a well may be like a crooked shaft in a mine, changing its direction from time to time. I found this to be the case in one into which I caused myself to be lowered in order to examine the internal structure of the glacier. For some time my descent was straight and direct, but at a depth of about fifty feet there was a landing-place, as it were, from which the opening continued its farther course at quite a different angle. It is within these cylindrical openings in the ice that those accumulations of sand collect which form the pyramids described above.

One may often trace the gradual formation of these wells, because, as they require certain similar conditions, they are very apt to be found in various stages of completion along the same track where these conditions occur. Fissures, for instance, will often be produced along the same line, because, as the mass of the glacier moves on, its upper portions, as they advance, come successively in contact with inequalities of the bottom, in consequence of which the ice is strained beyond its power of resistance and cracks across. Rivulets are also likely to be renewed summer after summer over the same track, because certain conditions of the surface of the glacier, to which I have not yet alluded, and which favor the more rapid melting of the ice, remain unchanged year after year. Of course, the wells do not remain stationary any more than any other feature of the glacier. They move on with the advancing mass of ice, and we consequently find the older ones considerably lower down than the more recent ones. In ascending such a track as I have described, along which fissures and rivulets are likely to occur, we may meet first with a sand-pyramid; at a certain distance above that there may be a circular opening filled to its brim with the sand which has just reached the surface of the ice; a little above may be an open well with the rivulet still pouring into it; or, higher up, we may meet an open fissure with the two semicircles opposite each other on the margins, but not yet united, as they will be presently by the closing of the fissure; or we may find near by another fissure, the edges of which are just beginning to wear in consequence of the action of the water. Thus, though we cannot trace the formation of such a cylindrical shaft in the glacier from the beginning to the end, we may, by combining the separate facts observed in a number, decipher their whole history.

In describing the surface of the glacier, I should not omit the shallow troughs, which I have called "meridian holes," from the accuracy with which they register the position of the sun. Here and there on the glacier there are patches of loose materials, dust, sand, pebbles, or gravel, accumulated by diminutive water-rills, and small enough to become heated during the day. They will, of course, be warmed first on their eastern side, then, still more powerfully, on their southern side, and in the afternoon with less force again on their western side, while the northern side will remain comparatively cool. Thus around more than half of their circumference they melt the ice in a semicircle, and the glacier is covered with little crescent-shaped troughs of this description, with a steep wall on one side and a shallow one on the other, and a little heap of loose materials in the bottom. They are the sun-dials of the glacier, recording the hour by the advance of the sun's rays upon them.

In recapitulating the results of my glacial experience, even in so condensed a form as that in which I intend to present them here, I shall be obliged to enter somewhat into personal narration, though at the risk of repeating what has been already told by the companions of my excursions, some of whom wrote out in a more popular form the incidents of our daily life which could not be fitly introduced into my own record of scientific research. When I first began my investigations upon the glaciers, now more than twenty-five years ago, scarcely any measurements of their size or their motion had been made. One of my principal objects, therefore, was to ascertain the thickness of the mass of ice, generally supposed to be from eighty to a hundred feet, and even less. The first year I took with me a hundred feet of iron rods, (no easy matter, where it had to be transported to the upper part of a glacier on men's backs,) thinking to bore the glacier through and through. As well might I have tried to sound the ocean with a ten-fathom line. The following year I took two hundred feet of rods with me, and again I was foiled. Eventually I succeeded in carrying up a thousand feet of line, and satisfied myself, after many attempts, that this was about the average thickness of the glacier of the Aar, on which I was working.

I mention these failures, because they give some idea of the discouragements and difficulties which meet the investigator in any new field of research; and the student must remember, for his consolation under such disappointments, that his failures are almost as important to the cause of science and to those who follow him in the same road as his successes. It is much to know what we *cannot* do in any given direction,— the first step, indeed, toward the accomplishment of what we can do.

A like disappointment awaited me in my first attempt to ascertain by direct measurement the rate of motion in the glacier. Early observers had asserted that the glacier moved, but there had been no accurate demonstration of the fact, and so uniform is its general appearance from year to year that even the fact of its motion was denied by many. It is true that the progress of boulders had been watched; a mass of rock which had stood at a certain point on the glacier was found many feet below that point the following year; but the opponents of the theory insisted that it did not follow, because the mass of rock had moved, that therefore the mass of ice had moved with it. They believed that the boulder might have slid down for that distance. Neither did the occasional encroachment of the glaciers upon the valleys prove anything; it might be solely the effect of an unusual accumulation of snow in cold seasons. Here, then, was another question to be tested; and one of my first experiments was to plant stakes in the ice to ascertain whether they would change their position with reference to the sides of the valley or not. If the glacier moved, my stake must of course move with it; if it was stationary, my stakes would remain standing where I had placed them, and any advance of other objects upon the surface of the glacier would be proved to be due to their sliding, or to some motion of their own, and not to that of the mass of ice on which they rested. I found neither the one nor the other of my anticipated results; after a short time, all the stakes lay flat on the ice, and I learned nothing from my first series of experiments, except that the surface of the glacier is wasted annually for a depth of at least five feet, in consequence of which my rods had lost their support, and fallen down. Similar disappointment was experienced by my friend Escher upon the great glacier of Aletsch.

My failure, however, taught me to sink the next set of stakes ten or fifteen feet below the surface of the ice, instead of five; and the experiment was attended with happier results. A stake planted eighteen feet deep in the ice, and cut on a level with the surface of the glacier, in the summer of 1840, was found, on my return in the summer of 1841, to project seven feet, and in the beginning of September it showed ten feet above the surface. Before leaving the glacier, in September, 1841, I planted six stakes, at a certain distance from each other, in a straight line across the upper part of the glacier, taking care to have the position of all the stakes determined with reference to certain fixed points on the rocky walls of the valley. When I returned, the following year, all the stakes had advanced considerably, and the straight line had changed to a crescent, the central rods having moved forward much faster than those nearer the sides, so that not only was the advance of the glacier clearly demonstrated, but also the fact that its middle portion moved faster than its margins. This furnished the first accurate data on record concerning the average movement of the glacier during the greater part of one year. In 1842 I caused a trigonometric survey of the whole glacier of the Aar to be made, and several lines across its whole width were staked and determined with reference to the sides of the valley;* for a number of successive years the survey was repeated, and furnished the numerous data concerning the motion of the glacier which I have published. I shall probably never have an opportunity of repeating these experiments, and examining anew the condition of the glacier of the Aar; but, as all the measurements were taken with reference to certain fixed points recorded upon the map mentioned in the note, it would be easy to renew them over the same locality, and to make a direct comparison with my first results after an interval of a quarter of a century. Such a comparison would be very valuable to science, as showing any change in the condition of the glacier, its rate of motion, etc., since the time my survey was made.

These observations not only determined the fact of the motion of the glacier itself, as well as the inequality of its motion in different parts, but explained also a variety of phenomena indirectly connected with it. Among these were the position and direction of the crevasses, those gaping fissures of unknown depths, sometimes

* All the trigonometrical measurements connected with my experiments were very ably conducted by Mr. Wild, now Professor at the Federal Polytechnic School in Zürich; they are recorded in the topographical survey and map of the glacier of the Aar, accompanying my "Système Glaciaire."

a mile or more in length, and often measuring several hundred feet in width, the terror, not only of the ordinary traveller, but of the most experienced mountaineers. There is a variety of such crevasses upon the glacier, but the most numerous and dangerous are the transverse and lateral ones. The transverse ones were readily accounted for after the motion of the glacier was admitted; they must take place, whenever, in consequence of the advance of the glacier over inequalities or steeper parts of its bed, the tension of the mass was so great that the cohesion of the particles was overcome, and the ice consequently rent apart. This would be especially the case wherever some steep angle in the bottom over which it moved presented an obstacle to the even advance of the mass. But the position of the lateral ones was not so easily understood. They are especially apt to occur wherever a promontory of rock juts out into the glacier; and, when fresh, they usually slant obliquely upward, trending from the prominent wall toward the head of the glacier, while, when old, on the contrary, they turn downward, so that the crevasses around such a promontory are often arranged in the shape of a spread fan, diverging from it in different directions. When the movement of the glacier was fully understood, however, it became evident, that, in its effort to force itself around the promontory, the ice was violently torn apart, and that the rent must take place in a direction at right angles with that in which the mass was moving. If the mass be moving inward and downward, the direction of the rent must be obliquely upward. As now the mass continues to advance, the crevasses must advance with it; and, as it moves more rapidly toward the middle than on the margins, that end of the crevasse which is farthest removed from the projecting rock must move more rapidly also; the consequence is, that all the older lateral crevasses, after a certain time, point downward, while the fresh ones point upward.

Not only does the glacier collect a variety of foreign materials on its upper surface, but its sides as well as its lower surface are studded with boulders, stones, pebbles, sand, coarse and fine gravel, so that it forms in reality a gigantic rasp, with sides hundreds of feet deep, and a surface thousands of feet wide and many miles in length, grinding over the bottom and along the walls between which it moves, polishing, grooving, and scratching them as it passes onward. One who is familiar with the track of this mighty engine will recognize at once where the large boulders have hollowed out their deeper furrows, where small pebbles have drawn their finer marks, where the stones with angular edges have left their sharp scratches, where sand and gravel have rubbed and smoothed the rocky surface, and left it bright and polished as if it came from the hand of the marble-worker. These marks are not to be mistaken by any one who has carefully observed them; the scratches, furrows, grooves, are always rectilinear, trending in the direction in which the glacier is moving, and most distinct on that side of the surface-inequalities facing the direction of the moving mass, while the lee-side remains mostly untouched.

It may be asked, how it is known that the glacier carries this powerful apparatus on its sides and bottom, when they are hidden from sight. I answer, that we might determine the fact theoretically from certain known conditions respecting the conformation of the glacier, to which I shall allude presently; but we need not resort to this kind of evidence, since we have ocular demonstration of the truth. Here and there on the sides of the glacier it is possible to penetrate between the walls and the ice to a great depth, and even to follow such a gap to the very bottom of the valley; and everywhere do we find the surface of the ice fretted, as I have described it, with stones of every size, from the pebble to the boulder, and also with sand and gravel of all sorts, from the coarsest grain to the finest; and these materials, more or less firmly set in the ice, form the grating surface with which, in its onward movement down the Alpine valleys, it leaves everywhere unmistakable traces of its passage.

We come now to the moraines, those walls of loose materials built by the glaciers themselves along their road. They have been divided into three classes, namely, lateral, medial, and terminal moraines. Let us look first at the lateral ones; and to understand them we must examine the conformation of the glacier below the *névé*, where it assumes the character of pure compact ice. We have seen that the fields of snow, where the glaciers have their origin, are level, and that lower down, where these masses of snow begin to descend toward the narrower valley, they follow its trough-like shape, sinking toward the centre and sloping upward against the sides, so that the surface of the glacier, about the region of the *névé* is slightly concave. But lower down in the glacier proper, where it is completely transformed into ice, its surface becomes convex, for the following reason. The rocky walls of the valley, as they approach the plain, partake of its high temperature. They become heated by the sun during the day in summer, so that the margins of the glacier melt rapidly in contact with them. In consequence of this, there is always in the lower part of the glacier a broad depression between the ice and the rocky walls, while, as this effect is not felt in the centre of the glacier, it there retains a higher level. The natural result of this is a convex surface, arching upward toward the middle, sinking toward the sides. It is in these broad, marginal depressions that the lateral moraines accumulate; masses of rock, stones, pebbles, dust, all the fragments, in short, which become loosened from the rocky walls above, fall into them, and it is a part of the materials so accumulated which gradually work their way downward between the ice and the walls, till the whole side of the glacier becomes studded with them. It is evident, that, when the glacier runs in a northerly or southerly direction, both the walls will be affected by the sun, one in the morning, the other in the afternoon, and in such a case the sides will be uniform, or nearly so. But when the trend of the valley is from east to west, or from west to east, the northern side only will feel the full force of the sun; and in such a case, only one side of the glacier will be convex in outline, while the other will remain nearly on a level with the middle. The large masses of loose materials which accumulate between the glacier and its rocky walls and upon its margins form the lateral moraines. These move most slowly, as the marginal portions of the glacier advance at a much slower rate than its centre.

The medial moraines arise in a different way, though they are directly connected with the lateral moraines. It often happens that two smaller glaciers unite, running into each other to form a larger one. Suppose two glaciers to be moving along two adjoining valleys, converging toward each other, and running in an easterly or westerly direction; at a certain point these two valleys open into a single valley, and here, of course, the two glaciers must meet, like two rivers rushing into a common bed. But as glaciers consist of a solid, and not a fluid, there will be no indiscriminate mingling of the two, and they will hold their course side by side. This being the case, the lateral moraine on the southern side of the northernmost glacier, and that on the northern side of the southernmost one, must meet in the centre of the combined glaciers. Such are the so-called medial moraines formed by the junction of two lateral ones. Sometimes a glacier may have a great number of tributaries, and in that case we may see several such moraines running in straight lines along its surface, all of which are called medial moraines in consequence of their origin midway between two combining glaciers. The glacier of the Aar, represented in the woodcut opposite, affords a striking example of a large

EXTERNAL APPEARANCE OF GLACIERS. 305

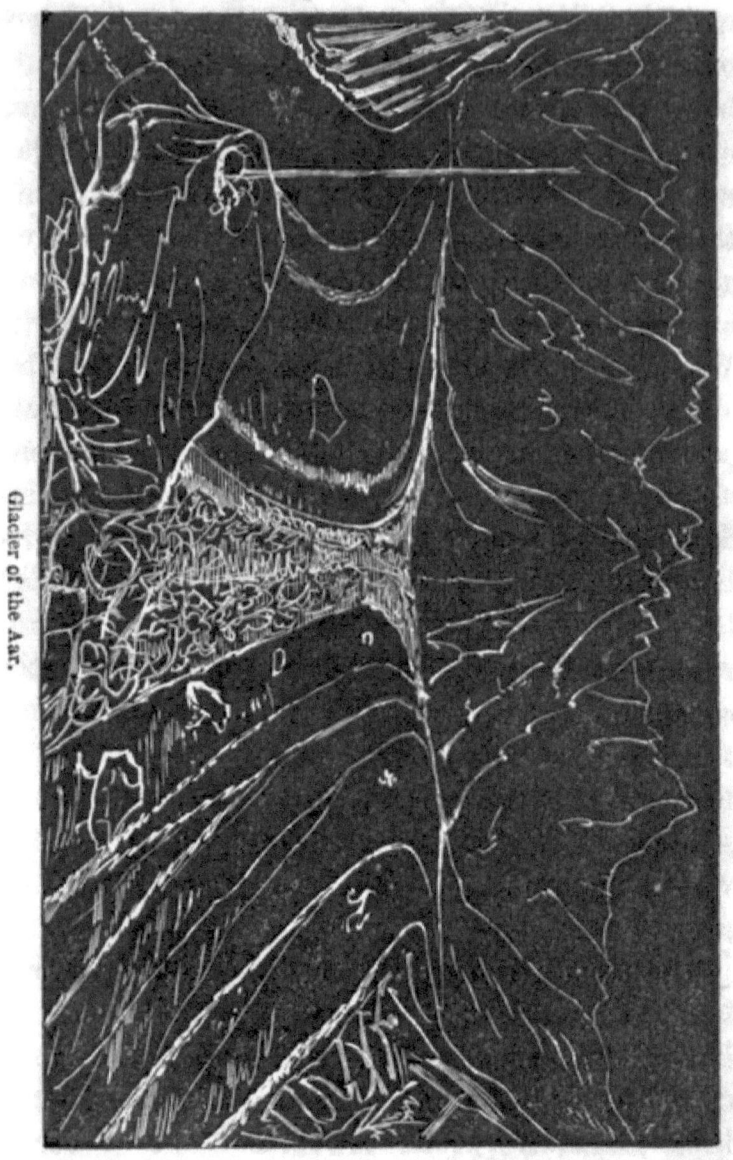

Glacier of the Aar.

NOTE.— The cuts on pp. 305 and 307 have accidentally been reversed,— what is on the left should be on the right, and *vice versa;* they must therefore be considered only as diagrams of a junction of two glaciers and of a terminal moraine.

T

medial moraine. It is formed by the junction of the glaciers of the Lauter-Aar, on the right-hand side of the wood-cut, and the Finster-Aar, on the left; and the union of their inner lateral moraines, in the centre of the diagram, forms the stony wall down the centre of the larger glacier, called its medial moraine. This moraine at some points is not less than sixty feet high. We have here an effect similar to that of the glacier-tables and the sand-pyramids. The wall protects the ice beneath it, and prevents it from sinking at the same rate as the surrounding surface, while its heated surface increases the melting of the adjacent surfaces of ice, thus forming longitudinal depressions along the medial moraines, in which the largest rivulets and the most conspicuous sand-pyramids, the deepest wells and the finest waterfalls, are usually met with. As the medial moraines rest upon that part of the glacier which moves fastest, they of course advance much more rapidly than the lateral moraines.

The terminal moraines consist of all the *débris* brought down by the glacier to its lower extremity. In consequence of the more rapid movement of the centre of the glacier, it always terminates in a semicircle at its lower end, where these materials collect, and the terminal moraines, of course, follow the outline of the glacier. The

EXTERNAL APPEARANCE OF GLACIERS. 307

wood-cut below represents the terminal moraine of the glacier of Viesch.

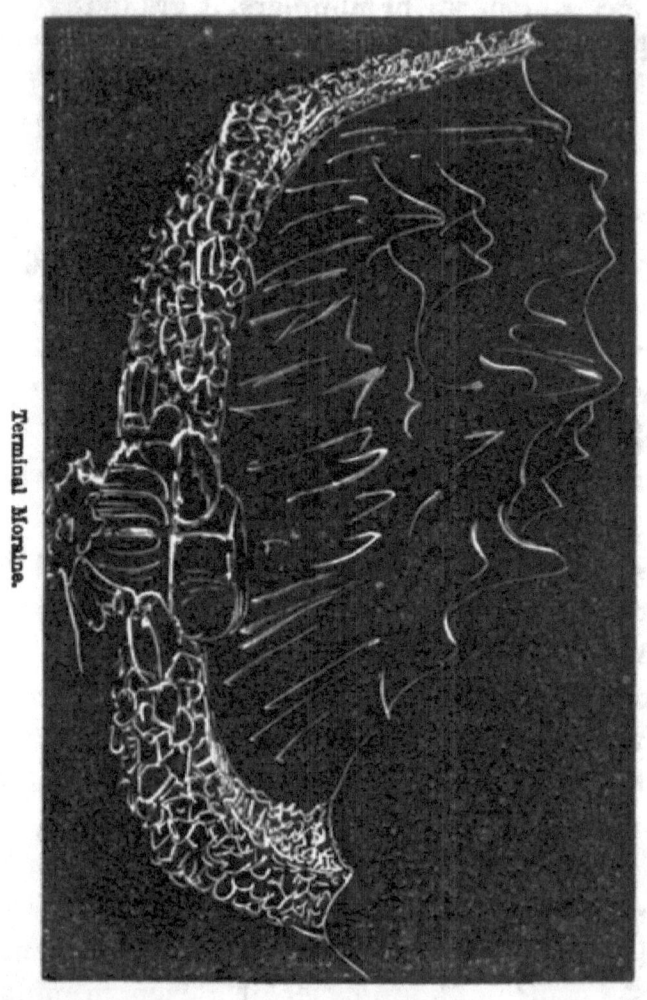

Sometimes, when a number of cold summers have succeeded each other, preventing the glacier from melting in proportion to its advance, the accumulation of materials at its terminus becomes very considerable; and when, in consequence of a succession of warm summers, it gradually melts and retreats from the line it has been occupying, a large semicircular wall is left, spanning the valley from side to side, through which the stream issuing from the glacier may be seen cutting its way. It is important to notice that such terminal moraines may actually span the whole width of a valley, from side to side, and be interrupted only where watercourses of sufficient power break through them. To suppose that such transverse walls of loose materials could be thrown across a valley by a river were to suppose that it could build dams across its bed while it is flowing. Such transverse or crescent-shaped moraines are everywhere the work of glaciers.

All these moraines are the landmarks, so to speak, by which we trace the height and extent, as well as the progress and retreat, of glaciers in former times. Suppose, for instance, that a glacier were to disappear entirely. For ages it has been a gigantic ice-raft, receiving all sorts of materials on its surface as it travelled onward, and bearing them along with it; while the hard particles of rock set in its lower surface have been polishing and fashioning the whole surface over which it extended. As it now melts, it drops its various burdens on the ground; boulders are the milestones marking the different stages of its journey, the terminal and lateral moraines are the framework which it erected around itself as it moved forward, and which define its boundaries centuries after it has vanished, while the scratches and furrows it has left on the surface below show the direction of its motion.

All the materials which reach the bottom of the glacier, and are moving under its weight, so far as they are not firmly set in the ice, must be pressed against one another, as well as against the rocky bottom, and will be rounded off, polished, and scratched, like the rock itself over which they pass. The pebbles or stones set fast in the ice will be thus polished and scratched, however, only over the surface exposed; but, as they may sometimes move in their socket, like a loosely mounted stone, the different surfaces may in turn undergo this process, and in the end all the loose materials under a glacier become more or less polished, scratched, and grooved. These marks exhibit also the peculiarity so characteristic of the grooves and scratches on the bed and walls of the valley; they are rectilinear, trending in the direction in which the superincumbent mass advances, though, of course, owing to the changes in the position of the pebbles or boulders, they may cross each other in every direction on their surface.

As the large materials are pressed onward with the finer ones, that is, with the sand, gravel, and mud accumulated at the bottom of the glacier, the component parts of this underlying bed of *débris* will be mixed together without any reference to their size or weight. The softest mud and finest sand may be in immediate contact with the bottom of the valley, while larger rocks and pebbles may be held in the ice above; or their position may be reversed, and the coarser materials may rest below, while the finer ones are pressed between them or overlying them. In short, the whole accumulation of loose *débris* under the glacier, resulting from the trituration of all kinds of angular fragments reaching the lower surface of the ice, presents a sort of paste, in which coarser and lighter materials are impacted without reference to bulk or weight. Those fragments which are most polished, rounded, grooved, or scratched, have travelled longest under the glacier, and are derived from the hardest rocks, which have resisted the general crushing and pounding for a longer time. The masses of rock on the upper surface of the glacier, on the contrary, are carried along on its back without undergoing any such friction. Lying side by side, or one above another, without being subject to pressure from the ice, they retain, both in the lateral and medial moraines, and even in the terminal moraines, their original size, their rough surfaces, and their angular form. Whenever, therefore, a glacier melts, it is evident that the lower materials will be found covered by the angular surface-materials now brought into immediate contact with the former in consequence of the disappearance of the intervening ice. The most careful observations and surveys have shown this everywhere to be the case; wherever a large tract of glacier has disappeared, the moraines, with their large angular boulders, are found resting upon this bottom layer of rounded materials, scattered through a paste of mud and sand.

We shall see hereafter how far we can follow these traces, and what they tell us of the past history of glaciers, and of the changes the climates of our globe have undergone.

THE END.

Cambridge: Stereotyped and Printed by Welch, Bigelow, & Co.

www.ingramcontent.com/pod-product-compliance
Lightning Source LLC
Chambersburg PA
CBHW030802230426
43667CB00008B/1024